WILFRID HODGES

Logic

Second Edition

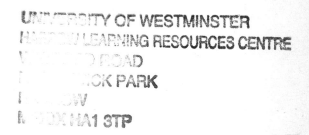

PENGUIN BOOKS

PENGUIN BOOKS

Published by the Penguin Group
Penguin Books Ltd, 80 Strand, London WC2R 0RL, England
Penguin Putnam Inc., 375 Hudson Street, New York, New York 10014, USA
Penguin Books Australia Ltd, 250 Camberwell Road, Camberwell, Victoria 3124, Australia
Penguin Books Canada Ltd, 10 Alcorn Avenue, Toronto, Ontario, Canada M4V 3B2
Penguin Books India (P) Ltd, 11 Community Centre, Panchsheel Park, New Delhi – 110 017, India
Penguin Books (NZ) Ltd, Cnr Rosedale and Airborne Roads, Albany, Auckland, New Zealand
Penguin Books (South Africa) (Pty) Ltd, 24 Sturdee Avenue, Rosebank 2196, South Africa

Penguin Books Ltd, Registered Offices: 80 Strand, London WC2R 0RL, England

www.penguin.com

First published in Pelican Books 1977
Reprinted in Penguin Books 1991
Second Edition published 2001
3

Set in 9.5/11.75 Adobe Minion
Typeset by Rowland Phototypesetting Ltd, Bury St Edmunds, Suffolk
Printed in England by Clays Ltd, St Ives plc

Contents

Acknowledgements

Nothing in this book is original, except perhaps by mistake. It is meant as an introduction to a well-established field of ideas; I only added a pattern and some examples. For discussions on various drafts of parts of the book, and related topics, I have many people to thank: chiefly David Wiggins, David K. Lewis, Ted Honderich, my wife Helen, and six generations of undergraduates specializing in various subjects at Bedford College. I thank Mariamne Luddington for her cheerful illustrations.

Permission to reproduce material in this book is acknowledged to the following sources:

To Pan Books Ltd for permission to quote in section 4 from Denys Parsons's collection of newspaper howlers, *Funny Amusing and Funny Amazing*, published in 1969.

To Harcourt Brace Jovanovich, Inc., and to Faber & Faber Ltd, for four lines from T. S. Eliot's 'Burnt Norton', from *Four Quartets*.

To the Oxford University Press for lines from 'Spelt from Sibyl's Leaves' by Gerard Manley Hopkins, from the fourth edition (1967) of *The Poems of Gerard Manley Hopkins*, edited by W. H. Gardner and N. H. Mackenzie, published by arrangement with the Society of Jesus.

To the Controller of Her Majesty's Stationery Office for an extract from paragraph 46 of *Corporal Punishment*, cmnd 1213, published in 1960, Crown copyright.

The extract from the Authorized Version of the Holy Bible – which is copyright – is used with permission.

Wilfrid Hodges
Bedford College
March 1976

Introduction

This book was written for people who want to learn some elementary logic, regardless of whether they are taking a course in it.

The book is written as a conversation between you (the reader) and me (the author). To keep the conversation from being too one-sided, I put in fairly frequent exercises. You are strongly urged to try to answer these as you reach them. Correct answers are given at the end of the book.

If you have a phobia of symbols, you should leave out the seven sections marked '+' in the text. These cover the branch of logic known as Formal Logic; they contain the worst of the mathematics.

English words and phrases which are being discussed are usually printed **bold**; most of these are listed in the index. A word is printed in quotes ' ' when we are chiefly interested in the way it occurs in some phrase.

For the second edition I rewrote sections 41 and 44 and added a few new exercises. I also corrected some typos and stupidities (and I thank the kind people who pointed these out). My special thanks to Jamal Ouhalla and Trevor Toube for helping me with linguistics and with chemical terminology respectively.

Wilfrid Hodges
Queen Mary
University of London
August 2000

Consistency

Logic can be defined as the study of consistent sets of beliefs; this will be our starting-point. Some people prefer to define logic as the study of valid arguments. Between them and us there is no real disagreement, as section 11 will show. But consistency makes an easier beginning.

1. Consistent Sets of Beliefs

Logic is about consistency – but not about all types of consistency. For example, if a man supports Arsenal one day and Spurs the next, then he is fickle but not necessarily illogical. If the legal system makes divorce easy for the rich but hard and humiliating for the poor, then it is unjust but not illogical. If a woman slaps her children for telling lies, and then tells lies herself, she may be two-faced but she need not be illogical.

The type of consistency which concerns logicians is not loyalty or justice or sincerity; it is *compatibility of beliefs*. A set of beliefs is consistent if the beliefs are compatible with each other. To give a slightly more precise definition, which will guide us through the rest of this book: a set of beliefs is called *consistent* if these beliefs could all be true together in some possible situation. The set of beliefs is called *inconsistent* if there is no possible situation in which all the beliefs are true.

For example, suppose a man believes:

> It would be wrong to censor violent programmes on television, because people's behaviour isn't really affected by what they see on the screen. All the same it would be a good idea to have more programmes showing the good sides of our national way of life, because it would straighten out some of the people who are always knocking our country. *1.1*

These beliefs are inconsistent: if it's really true that people's behaviour is not affected by what they see on the television screen, then it can't also be true that critics of a country will be reformed by what they see on the television screen. There is no possible situation in which (1.1) could all be true.

Inconsistency of beliefs is not at all the same thing as stupidity or unreasonableness. Take for example the man who believes:

> During the last five years I have been involved in three major accidents and several minor ones, while driving my car. After two of the major accidents, courts held me responsible. But basically I'm a thoroughly safe driver; I've simply had a run of bad luck. *1.2*

This man is almost certainly deceiving himself when he says he is a safe driver. His views are unreasonable. But they are not inconsistent: there is a possible (but unlikely) situation in which all of (1.2) would be true.

Or consider the man who believes:

> The surface of the earth is flat (apart from mountains, oceans and other relatively small bumps and dips). When people think they have sailed round the earth, all they have really done is to set out from one place and finish up in another place exactly like the one they started from, but several thousand miles away. *1.3*

This could all have been true, if the universe had been different from the way we know it is. A flat earth like the one described is a possibility. What is *not* a possibility is a flat earth with properties like the real earth as we know it. The beliefs in (1.3) are consistent in themselves, even though they are not consistent with the known facts.

Our definition of consistency can be applied to beliefs one at a time too. A single belief is called *consistent* if it could be true in some possible situation. An inconsistent belief is said to be *self-contradictory* and a *contradiction.*

For example, suppose a man tells you:

> I have invented an amazing new sedative which makes people faster and more excited. *1.4*

Then you can tell him – if you feel it would help – that he believes something self-contradictory. There is no possible situation in which a thing that made people faster and more excited could also be a sedative.

Is consistency of beliefs a virtue? Is it something we should spend time trying to achieve?

To some extent this is a question in a vacuum. People are never *deliberately* inconsistent in their beliefs. It is simply impossible to believe, fully and without reservation, two things which you know are inconsistent with each other.

Exercise 1A. You know that human beings normally have two legs. Try to convince yourself that they normally have five. (Allow yourself at least a minute.)

It seems we are obliged to believe only what we think is consistent, without having any real choice in the matter. In this way we are all logicians, simply because we are human. When we study logic, we are teaching ourselves to do deliberately, by rule, something we have been doing semi-consciously, by hunch, ever since the age of four.

Exercise 1B. Test your hunches by deciding which of the following sets of beliefs are consistent. (At this stage, don't take any of the examples too solemnly.)

1. I've never drawn anything in my life. But if I sat down to it now, it would take me two minutes to produce a drawing worth as much as anything by Picasso.
2. I knew I would never get pregnant. But somehow it just happened.
3. There is no housing shortage in Lincoln today – just a rumour that is put about by people who have nowhere to live.
4. Walter joined the friendly club two years ago, and has been one of its most loyal members ever since. Last year he paid for the holidays of precisely those club members who didn't pay for their own holidays.
5. The handless hold the hoe.
   A pedestrian walks, riding on a water buffalo.
   A man passes over the bridge;
   The bridge but not the water flows.

Expressing Beliefs in Sentences

Logic, then, is about beliefs and about when they are consistent with each other. But beliefs are hard to study directly: they are invisible, inaudible, weightless and without perceptible odour. To ease our task, we shall think of beliefs as being expressed by *written sentences*. This is not as straightforward as one might think. Two people can write down the same sentence and mean entirely different things by it. T. S. Eliot's despairing cry is often quoted:

> Words strain,
> Crack and sometimes break, under the burden,
> Under the tension, slip, slide, perish,
> Decay with imprecision, will not stay in place,
> Will not stay still.
>
> 'Burnt Norton', V

This is why we must spend a few sections examining how words are related to beliefs.

2. Beliefs and Words

When people want to let other people know what their beliefs are, they put them into words. We shall take a cue from this: instead of studying beliefs directly, we shall study the sentences which are used to state them. There are hazards in this.

In the first place, *many sentences do not naturally state beliefs*. For example, questions and commands don't express beliefs, at least not directly. (True, a person may betray his beliefs by the questions he asks.)

We shall limit ourselves to a class of sentences which are particularly suited to stating beliefs. These sentences are the so-called *declarative* sentences, and section 3 will explain what they are.

Second, *one sentence may have two different meanings*; in other words, it may be *ambiguous*. An ambiguous sentence may express either of two quite different beliefs. In section 4 we shall consider ambiguity, its types and its implications.

Third, *one sentence can be used on different occasions to talk about different things*. The weather forecaster who says 'It's going to rain tomorrow' is talking about the day after the day on which she makes her prediction. If she makes the same prophecy on two hundred different days in one year, she is using the same sentence to express two hundred different beliefs. A sentence may express a true belief when it is used in one situation, and an incorrect belief in another situation.

In section 5 we shall examine how the situation in which a sentence is used determines what it is about. This forms the theory of *reference*, which is a vital part of logic – unlike ambiguity, which is just a nuisance.

Fourth, *it may not always be clear whether a given sentence does correctly state a given belief*. This happens with confused or inchoate beliefs. It also happens because there is a nasty borderline between telling the truth in a misleading way, and telling downright lies. Like ambiguity, this is a tiresome nuisance; section 7 will consider it.

Fifth, since this is a book, all sentences in it will be written down. But *the written word is not a perfect substitute for the spoken word* – all the nuances of stress and intonation are lost.

One final difference between beliefs and sentences is that *sentences have to be in a language*. In this book, nearly all sentences are in English. Beliefs themselves are not in any language – in fact animals which speak no language can still believe things. It has been maintained that logicians have been unduly influenced by the grammar of English and other European languages. If the Inuit had invented the study of logic, it is said, the whole subject would have been quite different. This may be so, but it seems unlikely. For example, Japanese logicians do very much the same kind of work as white American ones.

3. Declarative Sentences

A declarative sentence of English is defined to be a grammatical English sentence which can be put in place of '*x*' in

Is it true that *x*? ***3.1***

so as to yield a grammatical English question. (Usually we shall omit the reference to English, since there is no danger of confusion with any other language.)

For example,

The price of beef has fallen. **3.2**

is a declarative sentence, since we can form the grammatical English question

Is it true that the price of beef has fallen? **3.3**

To apply the definition we have just given, we have to be able to tell the difference between grammatical and ungrammatical strings of words. This difference is important for logic, and in some ways very perplexing. So before we look for any further examples of declarative sentences, we shall make a detour to decide what 'grammatical' means.

The roots of grammar lie in the feelings, which every speaker of the language has, that certain strings of words are 'correctly put together' and others are not. These feelings are hard to describe and explain; but, for the most part, two speakers of the same language will agree about what feels right and what feels wrong. We call a string of words *grammatical* if most speakers of the language would accept it as correctly formed.

Strings of words can feel wrong in different ways and to different extents. We shall taste a few samples.

SAMPLE I

These pages must not be removed or defaced. **3.4**

Your breast will not lie by the breast **3.5**
Of your beloved in sleep.

These are two perfectly grammatical sentences; it's hard to see anyone raising an objection to either of them.

SAMPLE II

*The very so not was, wasn't it? **3.6**
*echo foxtrot golf hotel **3.7**

(3.6) and (3.7) are disastrously wrong. By any reckoning they are ungrammatical. (Linguists put * at the beginning of a phrase to indicate that it's ungrammatical.)

SAMPLE III

*Did you be angry with Sally? **3.8**
*Her train was decorated with silver gorgeous lace. **3.9**
*Government hastening collapse of economy. **3.10**
*By swaggering could I never thrive. **3.11**

None of (3.8)–(3.11) are completely acceptable in standard English. If a foreigner said one of them, we would know at once what he meant, but we might well correct him. The natural corrections are, respectively,

Were you angry with Sally? **3.12**
Her train was decorated with gorgeous silver lace. **3.13**
The government is hastening the collapse of the economy. **3.14**
I could never thrive by swaggering. **3.15**

Of course there are many situations where a slightly ungrammatical sentence is more appropriate than a strictly grammatical one. For example, newspaper headlines (such as (3.10)) are meant to catch the eye and give a rapid indication of what's in the print below them. It would be silly to expect them to contain all the 'of's and 'the's. Likewise only a pedant with time to kill would think of correcting the grammar of people's shopping lists. Again, everyone allows poets to break the rules (for emphasis, for special effect, or just to make it scan).

It will be useful to have a term for the kind of mistake which occurs in (3.8)–(3.11). The linguists tell me they have no such term, so I steal one from the mathematicians. I shall say that one sentence is a *perturbation* of another if the first sentence is grammatically wrong, but so nearly right that if it was used by someone whose English was imperfect we could safely correct it to the second sentence. Thus (3.8) is a perturbation of (3.12).

SAMPLE IV

?The pianist then played a red hat topped with geraniums and wisdom. **3.16**
?He ate a slice of boredom. **3.17**
?Nothing spoke her more than a hot bath. **3.18**
?The civilization of the ancient Persians fervently knew at least two inches. **3.19**

(3.16)–(3.19) put concepts together in impossible ways, but otherwise they are sound. Some logicians regard sentences like these as contradictions, while others take the view that a sentence like (3.16) is too nonsensical to be regarded as stating any belief. Grammarians also differ among themselves about whether these sentences should be counted as grammatical. (The ? at the beginning indicates doubtful grammaticality.)

The kind of mistake which occurs in sentences like (3.16)–(3.19) is called a *selection violation.* (The term comes from a grammatical theory of Noam Chomsky, though it is a theory he no longer supports.) Selection violations are easy to recognize by their bizarre and poetic feel. In fact they play an important role in poetic or metaphorical writing. By committing a selection violation deliberately, a writer can force her prosaic readers to forget the literal sense of what she says; since they can make nothing of her words if they take them literally, they have to notice the colours and the overtones.

Exercise 3A. Examine each of the sentences below, and decide whether it is

(a) grammatically correct,
(b) a perturbation of a grammatically correct sentence (say which),
(c) a sentence committing a selection violation, or
(d) hopelessly ungrammatical.

1. Colourless green ideas sleep furiously.
2. Furiously sleep ideas green colourless.
3. Please pass me a butter.
4. I bet there's dozens of people you've forgotten to invite.
5. singing each morning out of each night my father moved through depths of height
6. My father moved through theys of we
7. You can't do nothing with nobody that doesn't want to win.
8. His face was calm and relaxed, like the face of an asleep child.
9. Not, Father, further do prolong
   Our necessary defeat.
10. Time and the bell have buried the day.

Now we can come back to our definition of declarative sentences. A string of words will be called a *declarative sentence* if it is a grammatical sentence which can be put in place of '*x*' in

Is it true that *x*? **_3.20_**

so as to yield a question which is grammatical in the fullest sense – i.e. not a perturbation, and not a sentence with a selection violation.

Thus a grammatical sentence telling part of a story or stating a scientific or domestic fact will normally be declarative:

The accused entered the hall by the front door. ***3.21***

The current through a capacitance becomes greater when the frequency is increased. ***3.22***

It could do with a spot more salt. ***3.23***

In contrast to these, *questions* are not declarative sentences:

When did you last see your father? ***3.24***
yielding
*Is it true that when did you last see your father?

Neither are *commands* or *invitations*:

Tell me what you really really want. ***3.25***

Neither, by our definition, are sentences containing selection violations:

?The heart consists of three syllables. ***3.26***

The following exercises are not just drill. They illustrate some more points about declarative sentences, and at least two of the answers are controversial.

Exercise 3B. Which of the following are declarative sentences?

1. Twice two is four.
2. Please write a specimen of your signature in the space provided.
3. Would you believe you're standing where Cromwell once stood?
4. That's true.
5. I promise not to peep.
6. Blackmail is wicked.

---

4. Ambiguity

A string of words is said to be *ambiguous* if it can be understood as a meaningful sentence in two or more different ways.

Broadly speaking, there are two kinds of ambiguity – though they often occur together. The first kind is *lexical ambiguity*; this occurs when a single word in the string can be understood in more than one way. For example, the string

I thought it was rum. ***4.1***

contains a lexical ambiguity, because the word **rum** can mean either 'strangely sinister' or 'a drink made from fermented molasses'. (The context would usually make clear which is meant, of course: but the string itself is ambiguous.)

The cobbler who advertised

Our shoes are guaranteed to give you a fit. ***4.2***

committed another lexical ambiguity.

The second type of ambiguity is *structural ambiguity*; this occurs when the words in the string can be grouped together in different ways. For example, the string

I heard about him at school. ***4.3***

is structurally ambiguous: does it say that I was at school when I heard about him, or that I heard about what he did when he was at school? A similar structural ambiguity occurs in

Tariq is a Persian carpet importer. ***4.4***

One kind of structural ambiguity deserves a special name. This is *ambiguity of cross-reference*: it occurs when a word or phrase in the string refers back to something mentioned elsewhere, but it isn't clear which thing. In the sentence

Tamsin's stuck at home with a migraine, but Diane said she'd be dealing with the Kaplan file today. ***4.5***

are we being told that the Kaplan file is in the safe hands of Diane, or that Tamsin's migraine is delaying it? In other words, who is 'she'? Different languages have different ways of keeping this kind of ambiguity under control; for example in spoken English a slight emphasis on 'she'd' makes it more likely that 'she' is Diane. But ambiguities still occur and cause trouble.

Exercise 4A. What kinds of ambiguity occur in the following?

1. Launching the ship with impressive ceremony, the Admiral's lovely daughter smashed a bottle of champagne over her stern as she slid gracefully down the slipways.
2. Miss Crichton pluckily extinguished the blaze while Herr Eckold pulled the orchestra through a difficult passage.
3. A seventeen-year-old Copnor youth was remanded in custody to Portsmouth Quarter Sessions by Portsmouth Magistrates yesterday after he had admitted stealing three bicycles, a record player, thirty-one records, a National Insurance Card, and two cases of false pretences.
4. The font so generously presented by Mrs Smith will be set in position at the East end of the Church. Babies may now be baptized at both ends.
5. The visit has been planned in the atmosphere of the general economy going on in the country. The lunch has been cut to the bare bones.
6. The Government were strongly urged to take steps to put a stop to the growing evil of methylated spirit drinking by the Liverpool justices at their quarterly meetings.

There are two ways of talking about ambiguity: we can talk of one string representing two different sentences, or we can say that one sentence has two different meanings. The first style is probably preferable, but no harm will accrue if we sometimes talk in the second way.

It is easy to generate bogus inconsistencies by ignoring lexical ambiguities. For example, some people said that J. M. Keynes's *General Theory of Employment, Interest and Money* was inconsistent, because it maintained that the level of savings is always equal to the level of investment, although Keynes admitted that if people saved all their money rather than spending it, there would be no investment. The ambiguity lies in the word **save**: in common parlance it means hoarding, but it has quite a different sense in Keynes's theory.

In cases like this, it may help to distinguish two different words which happen to be written alike. Thus we can distinguish **saving**$_1$, which means hoarding, from Keynes's technical term **saving**$_2$, which means something more subtle.

Structural ambiguities may also lead to spurious inconsistencies. The best defence is to rewrite the offending sentences so as to remove the ambiguity.

Exercise 4B. Each of the following strings contains a structural ambiguity. Rewrite each string in two different ways, neither of them ambiguous, to show two possible interpretations of the string.

1. I shall wear no clothes to distinguish me from my fellow citizens.
2. He only relaxes on Sundays.
3. He gave each guest a glass of rum or gin and tonic.
4. Most of the nation's assets are in the hands of just one person.
5. Dogs must be carried. (*Sign by the escalators in London Underground stations.*)

When is a Sentence True?

Most contemporary logicians believe that there are two fundamental links between words and things. The first link is that declarative sentences are *true in certain situations*, and *not true in other situations*. The second link is that certain phrases *refer to* things in certain situations. Of course science advances, and it may be that some Einstein of logic will appear in A.D. 2050 and convince us that some quite different and hitherto unimagined notion is the key to the relations between words and things. But at present there is no sign of this.

Perhaps the first thinkers to take seriously the questions we now consider were the Heraclitean philosophers of ancient Greece, who maintained that 'It is impossible to say anything true about things which change.' One of them, Cratylus, found the whole matter so distressing that he thought it best to stop talking altogether, and simply waggle his finger.

5. Truth and References

The weather forecaster, we recall, makes the same prediction on many different days. Sometimes she is right, sometimes she is wrong. This illustrates an important point: *one and the same sentence can be used to make a true statement in one situation and an untrue statement in another situation.*

When a person utters a sentence, various parts of the sentence refer to various things in the world. The things which are referred to will normally depend on who uttered the sentence, and where and when and how she uttered it – in short they will depend on the *situation* in which the sentence was uttered. There are rules which determine what things are referred to by which parts of the sentence. As we shall see, these rules are complicated, and it takes children several years to learn them.

The examples which follow are somewhat oversimplified. The thing referred to by a phrase in a situation is called the *reference* of that phrase.

(1) There are some words and phrases whose reference depends entirely on where and when they are uttered. For instance:

I, now, here, the future, this month, next door. *5.1*

We can also put words like **will** and **was** in this group, because they refer to the times after or before the time of utterance. (Words of this group are said to be *token-reflexive.*)

(2) Some words and phrases require the speaker to give some special indication of what is referred to. For example, if I say 'You' when there are several people in the room, then the usual convention is that 'You' refers to the person I'm looking at or pointing to. Some other phrases which need a pointing finger are

this chair, the house over there, thus. *5.2*

(3) A proper name, such as **Arthur**, is used to refer only to people called Arthur. But of course thousands of people are called Arthur. The convention is that if you and I are in a situation where only one person called Arthur is likely to come to mind, then I can use the name **Arthur** to refer to that one person.

(4) Some phrases, such as **the latter reason** or **the aforementioned reindeer**, have a reference which depends entirely on the reference of some previously used phrase. This is a phenomenon called *cross-referencing*; we shall meet it many times again. (Recall the ambiguities of cross-reference which we discussed in section 4.) A simple example is:

Brutus killed Caesar by stabbing **him.** *5.3*

Here **him** refers to whatever the word **Caesar** refers to; i.e. it refers to the man Caesar. Pronouns such as **he, him, it** are often used in this way.

(5) Some phrases have the same reference whatever the situation; for example

sulfur, eighty-two, the planet Jupiter, the emotion of anger. *5.4*

As these examples show, the rules are complicated. Note that **I** comes under (1), **you** comes under (2) and **he** comes under (4); this indicates that a lot more than grammar is involved in the rules. Also one word may change its reference several times during a conversation.

Exercise 5. The following passage is about a situation in which Jacob wrestled with a man. Find every occurrence of **he, him, I, me, thou, thee** in the passage, and say who (Jacob or the man) is the reference at each occurrence.

> And when he saw that he prevailed not against him, he touched the hollow of his thigh; and the hollow of Jacob's thigh was out of joint, as he wrestled with him. And he said, Let me go, for the day breaketh. And he said, I will not let thee go, except thou bless me. And he said unto him, What is thy name? And he said, Jacob. And he said, Thy name shall be called no more Jacob, but Israel: for as a prince hast thou power with God and with men, and hast prevailed. And Jacob asked him, and said, Tell me, I pray thee, thy name. And he said, Wherefore is it that thou dost ask after my name? And he blessed him there.
>
> *Genesis 32: 25–9, Authorized Version*

The truth or untruth of a sentence commonly depends on the references of its parts; these references may change from one situation to another, and this is why the sentence may be true in one situation but not in another.

Rules are sometimes broken. If I tell you

The key is in the box by the phone. **5.5**

when in fact there is no box by the phone, then the phrase **the box by the phone** has no reference. The sentence (5.5) is being used improperly, because a phrase which ought to have a reference has none; in such cases we shall say that *referential failure* occurs.

Carelessness about referential failure easily leads to mistakes in logic. A particularly glaring example occurs in an argument which St Anselm of Canterbury used in order to prove the existence of God. Anselm reckoned it would be enough if he could prove that 'that than which no greater thing can be conceived' must exist. In order to prove that this thing does exist, he argued along the following lines:

> Suppose that than which no greater thing can be conceived doesn't exist outside our minds. Then it is not as great as it would have been if it had existed. Therefore we can conceive something greater than that than which no greater thing can be conceived; which is impossible. Therefore our original supposition is incorrect. ***5.6***

Now in the second sentence of (5.6), Anselm is assuming that there is something for 'it' to refer to. In view of the cross-referencing, this means he is assuming that the expression **that than which no greater thing can be conceived** refers to something. But he is not entitled to assume this, because it's precisely what he was supposedly proving in this argument. This makes the argument completely unconvincing.

Anselm's disregard for referential failure is rather extreme. But people often use words in ways which fail to make clear what they are talking about. The psychologist Jean Piaget noticed that young children frequently do this; for example he heard an eight-year-old child explaining the workings of a tap as follows:

> That and that is that and that because there it is for the water to run through, and that you see them inside because the water can't run out. The water is there and cannot run.† ***5.7***

This should be counted as an example of referential failure; although the child himself knows what he means, he uses words in such a way that the conventions of language fail to provide the needed references. Piaget thought that young children believe that references pass magically from their minds into the minds of the people they are talking to. An alternative explanation is that young children have not yet learned all the complex linguistic and social rules which determine references.

6. Borderline Cases and Bizarre Situations

If a declarative sentence is not true in a certain situation, we shall say that it is *false* in that situation. For example, the sentence

> The King of France is bald. ***6.1***

† Jean Piaget, *The Language and Thought of the Child*, Routledge & Kegan Paul, 1959, p. 104f.

is not true in the present state of affairs, because there is a referential failure – France has no King. Therefore we shall count (6.1) as being false in the present situation. The sentence

The King of France is hairy. *6.2*

is also false, for the same reason.

Logicians say that the *truth-value* of a declarative sentence is Truth when the sentence is true, and Falsehood when it is false. In any situation, a declarative sentence has just one truth-value: either Truth or Falsehood. There is something in this to catch the imagination. Life seems full of half-truths, grey areas, borderline cases, but Logic stands with sword uplifted to divide the world cleanly into the True and the False.

> Let life, waned, ah let life wind
> Off her once skeined stained veined variety upon, all on two
> spools; part, pen, pack
> Now her all in two flocks, two folds – black, white; right, wrong;
> reckon but, reck but, mind
> But these two; ware of a world where but these two tell, each off
> the other . . .
>
> *from* Gerard Manley Hopkins, 'Spelt from Sibyl's Leaves'

Many people have been attracted to logic by some such feeling. But honest thinkers must ask themselves whether this clean and absolute division into Truth and Falsehood is perhaps no more than a verbal illusion. Maybe Truth itself has degrees and blurred edges?

We shall try to answer this question. But first spare a thought for Ted Bartlett, who got fatter as he grew older – as Figure 1 illustrates.

Notice that even when Ted was definitely fat, it was possible for him to get still fatter. There can be two really fat people, one of them fatter than the other. In the same way it may be that my jokes are really funny, and your jokes are really funny too, but yours are much funnier than mine. Adjectives like **fat** and **funny** which have this feature will be called *scaling adjectives*. All of the following adjectives are scaling:

happy, expensive, heavy, unpleasant. *6.3*

The adjective **straight** is not a scaling adjective. If you and I draw lines, and your line is straighter than mine, then mine can't really be straight. Likewise **silent** is not a scaling adjective: if your machine is more silent

Figure 1

than mine, then mine isn't really silent. Some other adjectives which are not scaling are

square, perfect, smooth, daily. *6.4*

Exercise 6. Which of the following adjectives are scaling?

1. cold
2. fast
3. circular
4. two-legged
5. old
6. red
7. free
8. accurate
9. generous
10. full

Is **true** a scaling adjective? It is not: if your statement is truer than mine, then mine is not wholly true. **More true** can only mean 'more nearly true' or 'nearer the truth'. In this sense there are no degrees of truth. Truth is absolute.

But this is not the end of the matter. There is another way in which truth can be inexact. Ted Bartlett is the key once more; the thing to notice this time is that there was no exact time at which he became fat. At twenty-six he definitely wasn't fat, at fifty-six he definitely was. But there was a period in the middle when one could only describe him by roundabout phrases like **not all that fat, really**. There is no precise cut-off point between fat and not fat. We express this by saying that **fat** has *borderline cases.*

Most adjectives have borderline cases. Obviously **funny** and the adjectives in (6.3) have them. But so do some non-scaling adjectives like **silent.** There are situations which could be described as silent or as not silent, depending on what you count as a noise. (In Norse mythology, Heimdallr can hear the wool grow on a sheep's back.) On the other hand it seems that **daily** has no borderline cases; there is a clear and exact dividing line between daily and not daily.

How does **true** fare? Unfortunately it fares as badly as it possibly could. To see this, consider Ted Bartlett yet again. This time, instead of asking whether he is fat, ask whether it is *true* that he is fat. The borderline area comes just where it did before.

Thus we see that although truth doesn't have degrees, it does have many borderline cases.

There is a paradox here. In a given situation a declarative sentence must be either true or false, and not both, as we saw; nevertheless there are situations in which a declarative sentence may be not definitely true and not definitely false, in fact indeterminate. The dividing sword cuts somewhere, but there may be no definite place where it cuts.

The implications for logic are quite serious. If **true** has borderline cases, then so does **consistent**. Consider the pair of sentences

> Ted Bartlett is not fat. ***6.5***
> Ted Bartlett's vital statistics are XYZ.

(6.5) may or may not be consistent, depending on what we put for XYZ. But we have seen that there are some XYZ which make it quite indeterminate whether (6.5) is consistent. We are forced to admit that *where borderline cases may arise, logic is not an exact science.*

The absoluteness of Truth receives some hard knocks from yet another quarter, in the shape of *bizarre situations*. The issue is this. When a person learns a language, she learns that certain words are appropriate for certain situations, and inappropriate for others. Sometimes she meets a new situation, which is so different from the ones in which she learned the use of a word that she simply can't say, on the basis of her previous experience with the language, whether the word is correct to use in this new situation. A child who knows that **bone** is the appropriate word for the hard parts of a roast chicken may well be unsure whether she can use it for the hard part of a plum.

An adult can usually find the answer by asking someone who knows the language better than he does. 'In French, can one use "compétent" of a butler?' But it sometimes happens that a situation is so new and unusual that no speaker of the language is equipped to say what words are appropriate for it. We shall call such situations *bizarre*.

Here is an example of a bizarre situation. The two cerebral hemispheres of the human brain are joined by a structure called the corpus callosum. Surgeons sometimes cut through the corpus callosum in order to control epilepsy. People whose corpus callosum has been severed are quite normal in most ways, but they show one or two very strange symptoms. Suppose we show an object to such a person, and then ask him to indicate what we showed him by writing a mark on a piece of paper. If the object fell in the left half of his field of vision, then he can answer our question with his left hand but not with his right. If the object fell in the right half, then he can answer with his right hand but not with his left. Does a person with a severed callosum, who can see a mouse to the left of him, *know* that he is seeing a mouse? It seems that he knows with one half of his brain but not with the other; but in this case is it appropriate to say simply that he *knows*?

In a bizarre situation it may be impossible to say whether or not a sentence is true – not because we are stupid or we lack the facts, but simply because our language is not sufficiently articulated.

In real life, bizarre situations are the exception. But we have already seen (in section 1, (1.3)) that in logic it may be necessary to consider imaginary situations, and anybody with a creative imagination can dream up any number of bizarre situations.

There is a moral for logicians. If you want to get definite answers, then avoid the bizarre. As far as possible, stick to matter-of-fact notions, and leave the flights of fancy to the philosophers.

7. Misleading Statements

There are some cases of doubtful truth-value which revolve round the interpretation of common English words like **and** or **all**. Since it would be quite impractical to try to avoid these words, we shall have to settle these cases by some rough–and–ready convention.

Four examples follow. In each case a person makes a misleading statement.

(1) A witness in the case of *Thumptmann v. Thumper* states that

> Mr Thumper hit Mr Thumptmann three times with the camera tripod, and Mr Thumptmann fell to the floor. *7.1*

What the witness actually saw was that Mr Thumptmann fell to the floor just before Mr Thumper came into the room, and Mr Thumper hit him three times with the camera tripod before he could get up.

(2) After the office party, a man admits to his wife

> I did kiss some of the girls. *7.2*

In fact he kissed all nineteen of them.

(3) After another office party, another man boasts to his wife

> All the girls kissed me. *7.3*

In fact there were no girls at the party.

(4) Shortly before his retirement presentation, a man says to his wife

> They're going to give me either a watch or a silver pot. *7.4*

In fact he knows he is going to be given both, but he hasn't yet made up his mind which to pawn.

Everybody will agree that these four statements are misleading. But are they true or not? One view is that none of these statements are true. On this view, the word **and** in a narrative implies **and then**, the word **some** implies **not all**, the word **all** implies **at least one**, and **either . . . or** implies **not both**. We shall say that this view puts the *strong reading* on the sentences (7.1)–(7.4).

There is another view. We can say that each of these men has told the truth but not the whole truth. On this view, each man has misled by omitting to mention something which any honest person would have mentioned, but not by saying something untrue. If we take this view, then we say that (7.1) strictly says nothing about the order of the two events described, even though people normally describe the earlier event before the later. Likewise we say that **some** is quite compatible with **all**, even though people don't normally say **some** when they could have said **all**. Similarly with the other two examples. We shall say that this latter view puts the *weak reading* on (7.1)–(7.4).

The question at issue here is not a practical or moral one; the four men are equally dishonest on either reading. But it is a question which affects logic.

To see this, imagine that the speaker in example (2) decides he had better make a clean breast of it:

> I did kiss some of the girls. In fact I kissed all of them – *7.5*
> nineteen there were.

Has the speaker of (7.5) contradicted himself? On the strong reading, he has; the second sentence is a correction of the first. On the weak reading he has not; the second sentence merely amplifies the first.

A choice is called for: must we adopt the weak or the strong reading? Arguments have been advanced on both sides, based on various theories about meaning. In this book we shall normally opt for the weak reading. We shall do so mainly because this has been the habit among recent logicians. The weak reading is usually much easier to describe than the strong one.

8. Possible Situations and Meanings

What situations are possible?

In common parlance, a thing is possible only if it is consistent with the known facts. A thing is possible if it could be so, given what we know; otherwise it is impossible. It is possible that Queen Elizabeth I was a virgin till her dying day; it is possible that there is a monster in Loch Ness; it is possible that I shall catch a cold tomorrow. On the other hand it is impossible for me to hold my breath for ten minutes, and it is not possible that the earth is flat.

But as we saw in section 1, this is not quite what a logician means when he talks of a possible situation. In logic, a situation is described as possible if it *could have been* the actual situation, forgetting what we know about the world. If things had worked out differently, I could have been a multimillionaire, I could have become a famous and successful disc-jockey – these are possible situations.

Again, the world could have been the way it actually was in 1066; the state of the world in 1066 is a possible situation.

Again, if evolution had worked out differently, I could have had compound eyes and six arms. If pigs had wings, maybe I could fly one over the Atlantic. All these are possible situations.

There are limits to what is possible. For example, there is no possible situation in which two plus two is anything but four. True, people could *count* differently; they could count

one, two, four, three, five, six, . . . ***8.1***

But if this was so, then two plus two would still be four, although it would be *called* three. In just the same way, two plus three is still five in Arabic countries, although Arab script writes five in a way which looks like our 0. We must remember that we are describing possible situations in *our* language, and not in the possible languages which people might adopt in those situations.

In logic, we are interested in possible situations only from the point of view of examining what is true in them. As we saw in section 6, if a situation is bizarre, it may be impossible to say whether or not a certain sentence is true in it. Even among possible situations, there are some which are too peculiar to serve any useful purpose in logic.

It is worth noting that we can refer to situations which are possible but not actual, without thereby committing a referential failure. For example,

I can refer to the state of Europe in the mid-seventeenth century, although that state no longer obtains. I can also refer to things and people that don't actually exist, provided I say something to indicate that I am talking about a possible situation in which they do exist. For example, I can tell you that

> In 1650 the Holy Roman Emperor still had certain powers of taxation. ***8.2***

without falling into referential failure, because the first words of (8.2) have shifted the situation back to 1650, when there was a Holy Roman Emperor. Phrases such as **in 1650** will be known as *situation-shifters*; we shall meet them again.

Philosophers, who like to be precise in their use of words, often use possible situations in order to explain subtle differences between the meanings of words. The method is to describe a possible situation in which one sentence would be true but another similar sentence would be false. For example, J. L. Austin explained the difference between **mistake** and **accident** by telling a short story:

> You have a donkey, so have I, and they graze in the same field. The day comes when I conceive a dislike for mine. I go to shoot it, draw a bead on it, fire . . . but as I do so, the beasts move, and to my horror yours falls.† ***8.3***

In the situation described by (8.3), I have shot your beast by accident, I have not shot it by mistake; which shows that shooting by mistake and shooting by accident are not the same thing.

The same method can be used to point up differences between legal terms. For example, to explain the difference between ownership and possession, I can point out that if I steal your watch and put it in my pocket, then I possess it but I do not own it.

A parent or a teacher uses this same method when she explains the difference between **cynical** and **sarcastic** by describing how one can be sarcastic without being cynical.

† J. L. Austin, *Philosophical Papers*, Oxford University Press, 1961, p. 133.

Exercise 8. For each of the following pairs of sentences, show that the two sentences mean different things, by describing a possible situation in which one is true and the other is false.

1. The alderman was lying.
   What the alderman said was untrue.
2. It's in your best interests to go to Corsica.
   It would do you good to go to Corsica.
3. He knows I'm at home.
   He thinks I'm at home, and I am.
4. Brutus killed Caesar.
   Brutus caused Caesar to die.
5. A crow is a kind of bird.
   The word 'crow' is used to denote a kind of bird.

In the last few examples, we have begun to talk of sentences being true in situations where the sentences are not even uttered. Obviously this makes sense; for example the sentence

> No human beings exist yet.

was true a hundred million years ago, when there was nobody around even to think it. In such cases, we use a sentence to describe a possible situation as if it was actual.

Is there any difference between possible situations and the actual situation, apart from the fact that only the last one is actual? Do possible but not actual situations exist in the real world? Questions like these have made many logicians deeply unhappy about possible situations. The whole notion seemed much too speculative and metaphysical to them. They hoped that logic would guide us to greater certainties, not to perplexing questions about imaginary states of affairs. Alas, perplexing questions do not go away if we ignore them. There may be some better approach to the links between language and the world; but is it likely that Cratylus (see p. 13) and his followers will lead us there?

Testing for Consistency and Validity

Our task is to determine when a set of declarative sentences is consistent. If the sentences are all short enough, then close inspection will probably give the answer. When the sentences are longer and more complex, we need rules to guide us. The tableau method, which we follow in this book, works by breaking the sentences down into smaller ones. Although the idea seems very obvious, it was first invented in the 1930s by the German mathematician Gerhardt Gentzen.

We shall see that the same method can be used to test the validity of arguments.

9. Consistent Sets of Short Sentences

We began this book by defining logic as the study of the consistency of sets of beliefs; we then saw that beliefs can be expressed by declarative sentences. Just as with beliefs, a set of declarative sentences is called *consistent* if there is some possible situation in which all the sentences are true. Henceforth our task is to determine when a set of declarative sentences is consistent.

Ideally, we should like a method which could be applied to any finite set of declarative sentences, and which was guaranteed to tell us whether or not the set is consistent. Such a method would constitute a *decision procedure* for consistency. In fact one can show mathematically that no such decision procedure could possibly exist. The best we can hope for is a method which will work efficiently in most of the cases we are likely to meet.

If the sentences in the set are all reasonably short – say five or six words apiece – then close inspection is probably the best method. A set of short sentences is not usually inconsistent unless it contains an inconsistent pair of sentences, and inconsistent pairs are for the most part easy to recognize.

For example, each of the following pairs is obviously inconsistent:

Schizophrenia is curable. ***9.1***
It's not true that schizophrenia is curable.

Little is known about Heimdalargaldr. ***9.2***
Plenty is known about Heimdalargaldr.

Hazel's dress was bright red all over. ***9.3***
Hazel's dress had broad blue stripes.

Occasionally we meet sets of short sentences which are inconsistent but contain no inconsistent pairs. The following example was devised by Lewis Carroll:

All puddings are nice. ***9.4***
This dish is a pudding.
No nice things are wholesome.
This dish is wholesome.

Note that if a set of sentences is inconsistent, then it remains inconsistent if we add more sentences to the set; this is the so-called *monotonicity* property of consistency. We can never remove a contradiction by adding a few irrelevant remarks (though we may be able to stop people noticing it).

Exercise 9A. Which of the following sets of sentences are consistent? (This exercise is extremely easy, and is chiefly meant to illustrate some ways of being inconsistent.)

1. That was hardly an adequate payment. That was an entirely adequate payment.
2. Matilda is a hen. Matilda has four legs. Julia is a hen.
3. Few people speak or understand Cornish. Cornish is similar to Breton. Cornish is very musical. Cornish is a widespread language.
4. His hat is quite different from yours. Your hat is just like mine. My hat is unique.
5. Alf is taller than Bernard. Bernard is taller than David. Alf is shorter than David.

6. Angela is younger than Chris. Diana is older than Brenda. Brenda is younger than Chris. Diana is older than Chris.

There are the inevitable controversial examples too. We have already seen that borderline cases and bizarre situations may cause trouble. A different kind of example is the problem of *moral conflict*. Consider the two sentences

I ought to do it. **9.5**
I oughtn't to do it.

It might seem obvious that (9.5) is inconsistent. But suppose for example that I have arranged to baby-sit for some friends, but just as I am setting out, my mother starts to feel ill. I ought to go to baby-sit, because my friends are relying on me; I ought not to go, because I should stay at home in case my mother needs help. Doesn't this show that both sentences of (9.5) can be true at once?

This example needs analysing in two steps. First, there is a possible structural ambiguity: does the 'not' (written as 'n't') go with the auxiliary verb 'ought' or with the phrase 'to do it'? We can paraphrase these two readings as:

I'm not under an obligation to do it. **9.6**
I'm under an obligation not to do it.

It seems clear that the correct reading is the second, in spite of the fact that the 'n't' is attached as a suffix to 'ought'.

So, moving to the second step of the analysis, (9.5) is roughly equivalent to the pair of sentences

I'm under an obligation to do it. **9.7**
I'm under an obligation not to do it.

The question whether (9.7) is inconsistent is a question about whether it's possible for us to be under incompatible obligations.

Exercise 9B. For each of the following pairs of sentences, decide whether the 'not' or 'n't' goes with the auxiliary verb 'can' etc. or with the phrase that follows. In the light of your answer, say whether the pair of sentences is consistent or not.

1. Lars can swim.
   Lars can't swim.
2. The hurricane may reach us tomorrow.
   The hurricane may not reach us tomorrow.
3. It could be glandular fever.
   It couldn't be glandular fever.
4. It could be glandular fever.
   It could not be glandular fever. (*spoken with stress on* 'not')
5. I did have sexual relations with that woman.
   I did not have sexual relations with that woman.

---

10. The Tableau Technique

Semantic tableaux, or *tableaux* as we shall call them for short, are a method for testing the consistency of sets of complex sentences. In outline, the method is as follows.

Suppose X is a set of complex sentences; we wish to test whether X is consistent. If it is consistent, then the sentences of X are true together in some situation. We try to describe such a situation, using sentences which are as short as possible. X itself forms a first attempt at a description, so we start by writing down X. This forms the beginning of the tableau.

We then take any complex sentence P in X, and we try to describe, using only sentences shorter than P, those situations in which P would be true. If, for example, P is true precisely when two shorter sentences Q and R are both true, then we add Q and R to the tableau:

On the other hand we may only be able to find two shorter sentences Q and R, such that P is true in precisely those situations where *at least one* of Q and R is true. There are then two possible ways of finding a situation in which X is all true; so we add the sentences Q and R to the tableau, but we make them branch out in different directions:

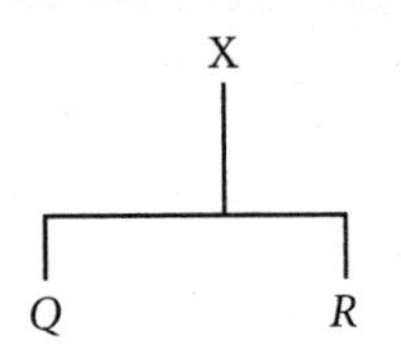

10.2

We then repeat the operation with another complex sentence, which may be Q or R or another sentence from X. We continue in this way for as long as we can.

At a stage half-way through the construction of the tableau, it may look something like this:

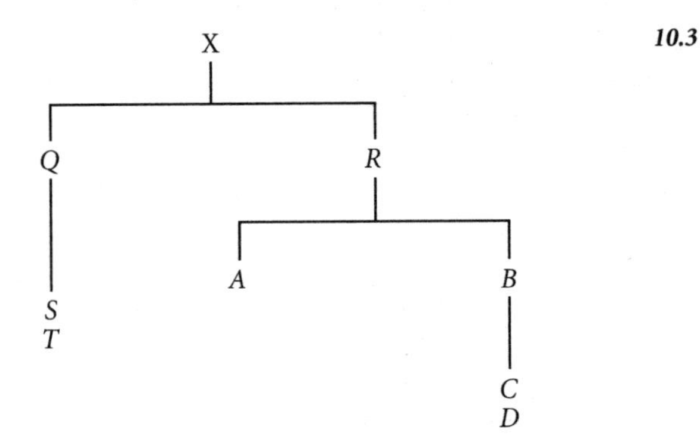

10.3

(Q, R, etc., are sentences.) We think of (10.3) as a tree with branches pointing downwards. In fact there are three branches in (10.3), namely

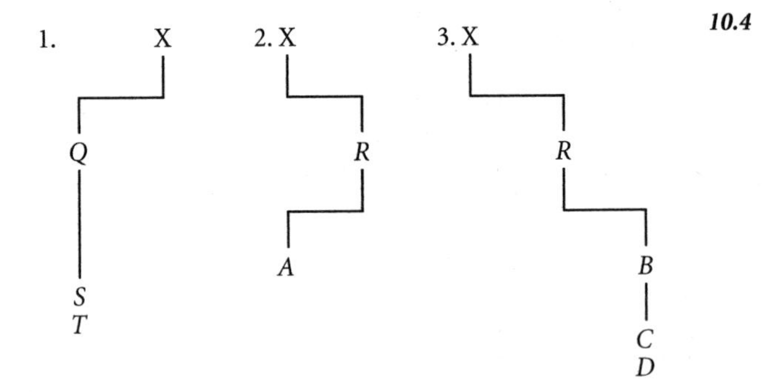

10.4

Each of these branches is an attempt to describe a possible situation. It succeeds in describing one if and only if the sentences in the branch form a consistent set; so we can extend each branch separately in just the same way as we extended the original set X.

When a branch contains an obvious inconsistency, it represents a failed attempt to describe a situation; we therefore *close* it by drawing a line across the bottom, and turn our attention to the other branches instead.

Eventually we shall reach a stage when it is impossible to extend any branch by adding new shorter sentences, and it is impossible to close off any more branches. The tableau is then finished. One of two things may happen.

(i) It may be that every branch of the finished tableau is closed; in this case we say that the tableau itself is *closed*. This means that every attempt to describe a situation in which X is true has led to contradictions. We can deduce that X is *inconsistent*.

(ii) It may be that there are some branches of the finished tableau which are not closed. Take just one of these branches – it doesn't matter which. Since the branch is not closed, the short sentences in it are not obviously inconsistent; in most cases we can see that they are in fact consistent. They then describe for us a situation in which X is true. Therefore X is *consistent*.

In case (i), the tableau gives us a proof that X is inconsistent; in case (ii), it usually shows that X is consistent.

We shall illustrate the tableau technique with two examples.

First example: is the following set of sentences (taken from the report of a chemical analysis) consistent?

> If cobalt but no nickel is present,
> a brown colour appears.
> Nickel and manganese are absent.
> Cobalt is present but only a green
> colour appears. ***10.5***

We begin the tableau by writing down (10.5). We choose any sentence of (10.5); say the second. This sentence is true precisely if 'Nickel is absent' and 'Manganese is absent' are both true. We therefore extend the tableau by writing

> If cobalt but no nickel is present,
> a brown colour appears. ***10.6***

✓ Nickel and manganese are absent.
Cobalt is present but only a
green colour appears.

|

Nickel is absent.
Manganese is absent.

(The tick shows that the second sentence has been dealt with.) Now we do the same, say with the third sentence:

If cobalt but no nickel is present,
a brown colour appears. ***10.7***
✓ Nickel and manganese are absent.
✓ Cobalt is present but only a
green colour appears.

|

Nickel is absent.
Manganese is absent.

|

Cobalt is present.
Only a green colour appears.

No contradiction has appeared yet among the shorter sentences; so we proceed to break down the first sentence. Now this sentence is true precisely if *either* a brown colour appears, *or* it's not true that 'Cobalt but no nickel is present'.† We therefore split the tableau into two branches:

✓ If cobalt but no nickel is present,
a brown colour appears. ***10.8***
✓ Nickel and manganese are absent.
✓ Cobalt is present but only a
green colour appears.

|

Nickel is absent.
Manganese is absent.

|

† In a situation where cobalt but no nickel is present, yet no brown colour appears, the sentence is false; also this is the only kind of situation which could make the sentence false. This point needs further discussion, and we shall come back to it in section 17 below.

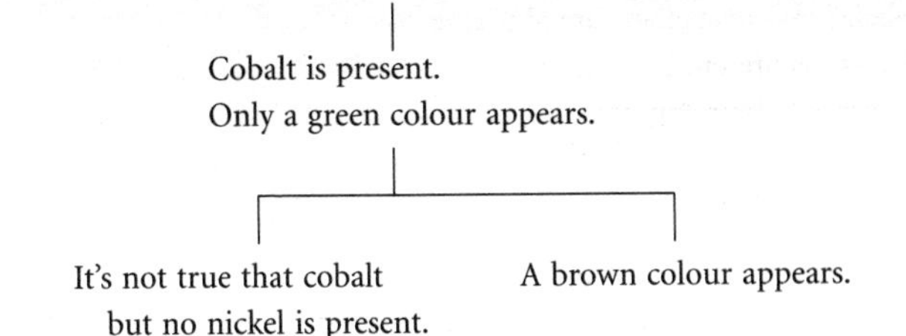

Cobalt is present.
Only a green colour appears.

It's not true that cobalt but no nickel is present. | A brown colour appears.

At this stage the right-hand branch contains two sentences, 'Only a green colour appears' and 'A brown colour appears', which obviously contradict each other. So we can close the right-hand branch:

10.9

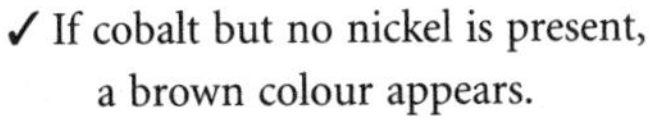

✓ If cobalt but no nickel is present, a brown colour appears.
✓ Nickel and manganese are absent.
✓ Cobalt is present but only a green colour appears.

Nickel is absent.
Manganese is absent.

Cobalt is present.
Only a green colour appears.

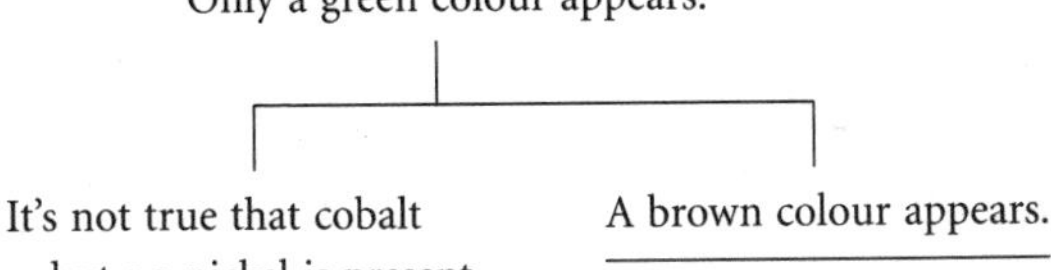

It's not true that cobalt but no nickel is present. | A brown colour appears.

The longest unanalysed sentence in (10.9) is the one at the foot of the left-hand branch. This sentence is true precisely if *either* cobalt is absent *or* nickel is present. We therefore extend the tableau below the left-hand branch:

10.10

✓ If cobalt but no nickel is present, a brown colour appears.
✓ Nickel and manganese are absent.
✓ Cobalt is present but only a green colour appears.

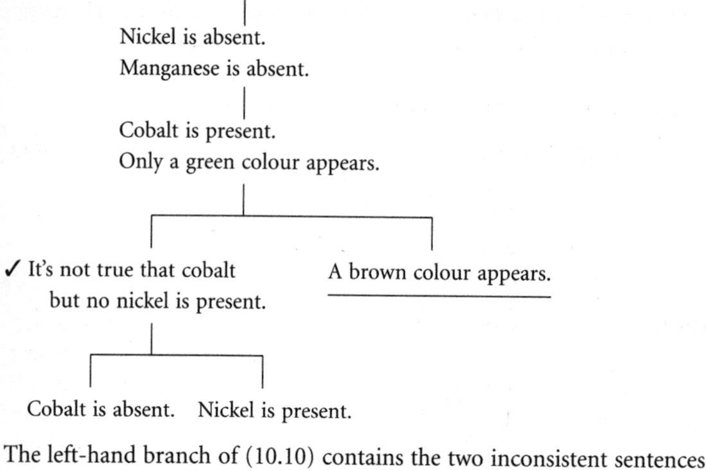

The left-hand branch of (10.10) contains the two inconsistent sentences 'Cobalt is present' and 'Cobalt is absent', so we can close it. The middle branch can be closed too, since it contains the inconsistent sentences 'Nickel is absent' and 'Nickel is present'. The resulting tableau is:

10.11

✓ If cobalt but no nickel is present,
a brown colour appears.
✓ Nickel and manganese are absent.
✓ Cobalt is present but only a
green colour appears.

|

Nickel is absent.
Manganese is absent.

|

Cobalt is present.
Only a green colour appears.

✓ It's not true that cobalt but no nickel is present.	A brown colour appears. (closed)

Cobalt is absent. (closed)	Nickel is present. (closed)

(10.11) is a finished tableau. All three branches are closed. Therefore (10.5) is *inconsistent.*

Second example: is the following set of sentences consistent?

> If cobalt but no nickel is present,
> a brown colour appears.
> Either nickel or manganese is absent.
> Cobalt is present but only a
> green colour appears. ***10.12***

The full tableau is as follows. You should work through this tableau step by step, to see why each sentence was added. Note in particular that where a sentence is in two branches which are not closed, then the shorter sentences got by analysing this sentence must be put at the foot of both branches. (Where does this apply in the tableau (10.13)?)

> ✓ If cobalt but no nickel is present,
> a brown colour appears.
> ✓ Either nickel or manganese is absent.
> ✓ Cobalt is present but only a
> green colour appears. ***10.13***

Nickel is absent.

Cobalt is present.
Only a green colour appears.

✓ It's not true that cobalt but no nickel is present.

Cobalt is absent.

Nickel is present.

A brown colour appears.

Manganese is absent.

Cobalt is present.
Only a green colour appears.

✓ It's not true that cobalt but no nickel is present.

Cobalt is absent.

Nickel is present.
⇑

A brown colour appears.

The tableau (10.13) is finished, but not all its branches are closed. We therefore pick one which is not closed; in fact there is just one unclosed branch, which is marked with an arrow. The short sentences in this branch are:

> Manganese is absent. Cobalt is present. Only a green colour appears. Nickel is present. ***10.14***

The sentences in (10.14) describe a situation, and in this situation the sentences (10.12) are true. Therefore (10.12) is *consistent.*

Exercise 10. Use tableaux to determine which of the following sets of sentences are consistent.

1. Mr Zak is a Russian spy.
   Mr Zak is not both a C.I.A. spy and a Russian spy.
   Mr Zak is a C.I.A. spy and a cad.
2. At least one of Auguste and Bruno lives in Bootle.
   At least one of Bruno and Chaim is an estate agent.
   Bruno is not an estate agent, and doesn't live in Bootle.
3. Either Yvonne or Zoe gave me this book last Tuesday.
   If Yvonne gave me this book, then I was in Oslo at Tuesday lunchtime.
   I was miles away from Oslo all Tuesday, and Zoe has never given me anything.

---

11. Arguments

Logic is sometimes defined as the study of valid arguments. What is an argument, and when is it valid?

An *argument,* in the sense that concerns us here, is what a person produces when he or she makes a statement and gives reasons for believing the statement. The statement itself is called the *conclusion* of the argument (though it can perfectly well come at the beginning); the stated reasons for believing the conclusion are called the *premises.* A person who presents or accepts an argument is said to *deduce* or *infer* its conclusion from its premises.

In logic books it's usual to write an argument with the premises first, then '*Therefore*', then the conclusion. For example:

> The car is making a hell of a noise, and it won't overtake properly. *Therefore* a gasket has blown. ***11.1***

Outside logic books, arguments occur in all sorts of forms. To pick an example off the shelf at random:

> There can, in fact, be no enduring solution to wretched living conditions unless the houses in which they are found are either altered or replaced. Until then, they will simply fill up again each time a family moves to something better. ***11.2***

In (11.2) the conclusion comprises the first sentence, and the second sentence indicates the premise. In logic book style:

> If houses in which wretched living conditions are found are neither altered nor replaced, they will simply fill up again each time a family moves to something better. *Therefore* there can be no enduring solution to wretched living conditions unless the houses in which they are found are either altered or replaced. ***11.3***

Notice that we had to repeat part of the conclusion in order to state the premise in full. People often present their arguments in an abbreviated form. In the next example, the conclusion is not stated outright at all:

> I think you ought to give it a rest. How would you like it if someone kept making jokes about *your* accent? ***11.4***

In logic book style, this might be:

> It's not pleasant to have people making constant jokes about one's accent. *Therefore* you ought to stop making jokes about so-and-so's accent. ***11.5***

Exercise 11. Rewrite each of the following as an argument in logic book style. (As you do them, notice the words and phrases like **in view of**, which serve to mark out the premise and the conclusion.)

1. Help is needed urgently, in view of the fact that two hundred people are dying every day.
2. When Communists operate as a minority group within unions, settlements by the established officials must be denounced as sellouts. It follows that strikes are unlikely to wither away in any democratic country so long as Communists have strong minority influence.
3. The nests of the verdin are surprisingly conspicuous, for they are usually placed at or near the end of a low branch.
4. The effect of ACTH on gout is not due to the increased renal uric acid clearance alone, since the effect of salicylates on this clearance is greater.
5. Some contribution to the magnetic field comes from electric currents in the upper atmosphere; otherwise we cannot account for the relation between the variations in the magnetic elements and the radiation received from the sun.

An argument is said to be *valid* if there is no possible situation in which its premises are all true and its conclusion is not true. An argument which is not valid is called *invalid*. When an argument is valid, its premises are said to *entail* its conclusion.

For example, this is a valid argument, so that its premises entail its conclusion:

> Charity workers earning at most £36 a week are exempt from council tax. I earn only £35 a week. I am a charity worker. *Therefore* I am exempt from council tax. ***11.6***

If the premises of (11.6) are true, then the conclusion must be true too; there is no other possibility. On the other hand the argument (11.1) is certainly not valid; a car could make a noise and lose power without having a broken gasket, even if a broken gasket is the most likely cause.

Notice that a valid argument need not have true premises. In fact I earn more than £35 a week; nevertheless (11.6) is still valid. People are notoriously prone to accept an argument as valid if they believe its premises and conclusion, and reject it as invalid if they disbelieve them. Take care.

If an argument consists of declarative sentences, then we can transform it into a certain set of declarative sentences which will be known as the *counterexample set*, as follows. The counterexample set consists of the premises, and the conclusion with the words 'It is not true that' tacked on to the front. For example, the counterexample set of (11.6) is

> Charity workers earning at most £36 a week are exempt from council tax. I earn only £35 a week. I am a charity worker. It is not true that I am exempt from council tax. ***11.7***

From the way in which we defined the validity of an argument, it's clear that *an argument is valid precisely if its counterexample set is inconsistent.*

This method of proving that an argument is valid – by showing that there is no possible situation in which its premises are true and its conclusion is false – is known as the method of *reductio ad absurdum* (the Latin for 'reduction to an absurdity'). For example, to prove (11.6) by *reductio ad absurdum*, we might argue:

> Assume that the premises are true and the conclusion is false. Since the first premise is true and the conclusion false, I can't be a charity worker earning at most £36 a week. But by the other two premises, I am a charity worker and I earn £35 a week, which has to be less than £36. This is absurd. Hence the argument (11.6) is valid. ***11.8***

This style of argument has caused some puzzlement, because it seems ridiculous to start the discussion by *assuming* something which is palpably untrue. A good answer is that when we say 'Assume the premises are true and the conclusion is false,' we are not asking anybody to *believe* any such thing; rather we are setting out to try to describe a possible situation in which the premises are true and the conclusion is false. When we reach an absurdity, this shows that there is no such possible situation.

All this was misunderstood rather badly by a nineteenth-century mathematician called James Smith, who thought he had proved that a certain number π is precisely 25/8. (It isn't.) He maintained that by an adapted version of *reductio ad absurdum*, it would be enough if he first assumed that π is 25/8, and then failed to deduce a contradiction from this assumption. Other mathematicians quickly pointed out that they could quite easily deduce contradictions from this assumption, but he simply took this as proof that they were incompetent.

If an argument expressed in declarative sentences is not valid, then we can show this by describing a possible situation in which its counterexample set is all true; such a situation is called a *counterexample* to the argument. For example, I can show that (11.1) is invalid by describing a car which makes a noise because its exhaust-pipe has a hole, and which won't overtake properly because the plugs are damp, though the gaskets are intact. There could be such a car; it provides a counterexample to (11.1).

We can test the validity of an argument by using a tableau to check the consistency of its counterexample set. If the tableau closes, then the counterexample set is inconsistent and so the argument was valid. If the tableau refuses to close, then some unclosed branch should describe for us a situation which is a counterexample to the argument.

For example, we test the argument:

> If Higgins was born in Bristol, then Higgins is not a Cockney. Higgins is either a Cockney or an impersonator. Higgins is not an impersonator. *Therefore* Higgins was born in Bristol. ***11.9***

The counterexample set of (11.9) is:

> If Higgins was born in Bristol, then Higgins is not a Cockney. Higgins is either a Cockney or an impersonator. Higgins is not an impersonator. It is not true that Higgins was born in Bristol. ***11.10***

We test the consistency of (11.10) by a tableau:

11.11

✓ If Higgins was born in Bristol,
then Higgins is not a Cockney.
✓ Higgins is either a Cockney or
an impersonator.
Higgins is not an impersonator.
It is not true that Higgins was
born in Bristol.

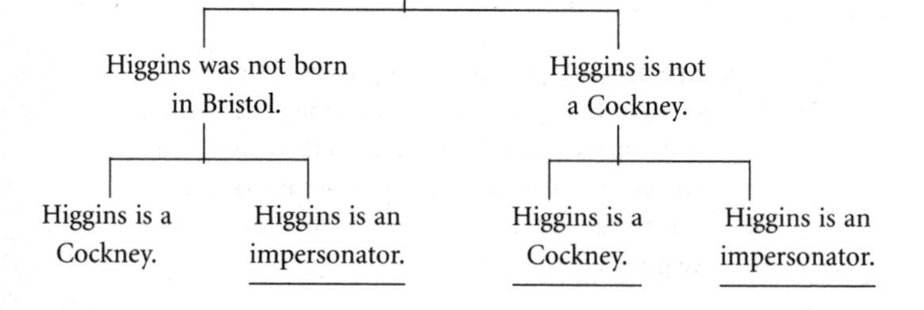

The left-most branch of (11.11) is not closed; it describes the situation

> Higgins is not an impersonator; he was not born in Bristol and he is a Cockney. ***11.12***

In the situation described by (11.12), the premises of the argument (11.9) are true and its conclusion is false. Therefore (11.12) forms a counter-example to the argument, and so the argument is *invalid.*

Closely related to valid arguments are *necessary truths*; these are declarative sentences which are true in every possible situation. Poem 449 in the *Oxford Book of Twentieth-Century English Verse* begins with a fine specimen:

> As we get older we do not get any younger. ***11.13***

Later we shall meet some less poetic examples.

Arguments can be good without being valid. We call an argument *rational* if its premises provide good reason for believing the conclusion, even if the reason is not absolutely decisive. For example, on the face of it (11.1) is quite rational – a blown gasket may be the most likely cause of noise and a loss of power, even if it is not the only possible cause.

If we can see that an argument is valid and has true premises, then we can see that its conclusion must be true too. Thus *a demonstrably valid argument whose premises are known to be true is rational.*

It would be pleasant to have some tests for the rationality of arguments, but this seems a vain hope. One difficulty is that the rationality of an argument depends on more than the stated premises; the evidence offered for the conclusion may become less convincing when further facts are pointed out. (With valid arguments this is not so: a valid argument remains valid even when new facts come to light.)

For example, a Government report on corporal punishment attacked those who believe that judicial corporal punishment would act as a deterrent:

> [These people] attribute the great increase in offences of violence against the person to the abolition of this penalty, but this argument overlooks the fact that there were similar increases before and after 1948, and that judicial corporal punishment was not available before 1948 for crimes of violence generally.† ***11.14***

† Para. 46 of *Corporal Punishment*, H.M.S.O., 1960.

The argument under attack here could be stated as follows:

> When judicial corporal punishment was abolished, offences of violence against the person increased. *Therefore* judicial corporal punishment acts as a deterrent against offences of violence against the person. ***11.15***

The argument (11.15) is attacked, not because it is unconvincing as it stands, but because it ceases to be convincing once we point out the known fact that offences of violence have increased in certain other circumstances too.

It seems, then, that in order to assess the rationality of an argument, we need to take into account all the known facts, and not just the stated premises. An argument is normally deployed against a background of known facts and agreed beliefs, and the rationality of the argument depends on what these facts and beliefs are. All this makes it hard to see how one could devise a simple and practical test of rationality.

Perhaps also we are to some extent free to choose for ourselves what we count as an adequate reason for believing a thing. By nature some people are more sceptical than others. In 1670 Chief Justice Vaughan put it rather well:

> A man cannot see by anothers eye, nor hear by anothers ear, no more can a man conclude or infer the thing to be resolved by anothers understanding or reasoning.

How are Complex Sentences Built Up?

Since our approach is to analyse sentences into their component parts, we ought to be sure we understand what the parts of a sentence really are. This demands a few sections on grammar.

There are two other reasons for having some grammar in a book about logic. First, the notions of structural ambiguity and scope, which a logician ought to know about, can only be understood with the help of a little grammar. Second, one of the techniques of modern logic is translation into certain formal languages; these languages are easier to set in motion if we gather up suitable grammatical ideas first.

12. Phrase-classes

Anybody who speaks a language can distinguish fairly accurately between the grammatical and the ungrammatical sentences of the language. In fact he can do something more: he can say which words within a grammatical sentence go together to form *natural groups*.

For example, any English speaker can tell you that in the sentence

> You can do anything but don't step on my blue suede shoes. ***12.1***

the following are all natural groups:

> You can do anything ***12.2***
> blue suede shoes
> don't step on my blue suede shoes

whereas the following are not:

do anything but *12.3*
You can do anything but don't step on my
my blue

This feeling for natural groupings is part of the raw data which a grammarian can use in constructing a grammar for a language. It's even less tangible than our feelings of what is and what is not grammatical. But it's a real feeling nevertheless, and people do agree to a remarkable extent about what the natural groups are, even if they can't say why.

The natural groups of words that occur in a language are called the *grammatical phrases* of the language. This is not a term with a precise meaning, because people do disagree about what groups are natural. But probably everybody will agree that the strings in (12.2) are grammatical phrases, while those in (12.3) are not. It will be convenient to count single words as grammatical phrases.

When a grammatical phrase occurs as a natural group in a sentence, it is called a *constituent* of the sentence. This is an important notion, and we shall use it frequently.

If a grammatical phrase occurs twice in a sentence, then the two occurrences are counted as *different constituents* of the sentence. For example, the phrase **blue suede shoes** forms two constituents of the following sentence:

My brother wanted **blue suede shoes** for Christmas, but I can't find a shop that sells **blue suede shoes**. *12.4*

It sometimes happens that a grammatical phrase occurs within a sentence, but doesn't form a natural group within the sentence. In this case the phrase does not count as a constituent of the sentence. For example, in

He wants a pair of dark blue suede shoes. *12.5*

'blue suede shoes' is not a natural group, since **dark blue** hangs closely together. The phrase **blue suede shoes** does *not* occur as a constituent of (12.5).

By convention, a whole sentence is counted as one of its own constituents.

Traditional English grammars group the grammatical phrases of English into *phrase-classes* (also known as *parts of speech*). Although the definitions given for these classes have often been absurd, these grammars agree very closely about what the main phrase-classes are. Four of the classes are called *noun, adjective, adverb* and *verb* respectively, containing words such as the following:

noun – **John, room, answer, play** ***12.6***
adjective – **happy, steady, new, large, round**
adverb – **steadily, completely, really, very**
verb – **search, grow, play, be, have, do**†

It's useful to have a test that determines when two phrases belong to the same class. The test we shall use is called the *frame test*. Don't regard it as any more than a rule of thumb. (In his book *Structural Linguistics* in 1947, Zellig Harris tried to use frames – which he called 'environments' – as a basis for the whole of linguistics.)

By a *frame* we mean a string of English words, among which the symbol '*x*' occurs once. '*x*' must occur only once, either between words or at the beginning or the end of the string. For example, here are three frames:

You really *x* Smith, don't you? ***12.7***
x people have heard of Xerxes. ***12.8***
Is it true that *x*? ***12.9***

The '*x*' serves to mark a hole in the frame, where other phrases can be put in.

If we take a frame and a grammatical phrase, and put the phrase into the frame in place of the '*x*', the result is a string of words. We say the frame *accepts* the phrase if this string is a grammatical sentence which has the introduced phrase as a constituent; if the frame doesn't accept the phrase, we say it *rejects* the phrase.

For example, the frame (12.7) accepts **hate**, but it rejects **hate Jack** and it rejects **my friend**. It accepts **hate**, because the sentence

You really **hate** Smith, don't you? ***12.10***

is grammatical and has **hate** as a constituent. (A single word occurring in a sentence always counts as a constituent of the sentence.) It rejects **hate Jack**, because although

You really **hate Jack** Smith, don't you? ***12.11***

† Randolph Quirk and Sidney Greenbaum, *A University Grammar of English*, Longman, 1973, p. 18.

is grammatical, **hate Jack** doesn't form a constituent of it. Finally it rejects **my friend** because

*You really **my friend** Smith, don't you? *12.12*

is not grammatical. (As in section 3, * means that the sentence is ungrammatical.)

Exercise 12. Write ticks and crosses in the following chart, to show which of the frames on the left accept which of the phrases at the top:

	health	a wife	wife	cameras	two cameras
He wanted to have *x*.					
He wanted to have the *x*.					
He wanted to have more *x*.					

Each frame picks out for us a class of grammatical phrases, namely those which it accepts. For example, the class of grammatical phrases accepted by the frame

Is it true that *x*? *12.13*

consists of the declarative sentences (see p. 5).

It often happens that two frames accept very nearly the same grammatical phrases, but not quite. For example, take the two frames

x will be mentioned afterwards. *12.14*
There will be a chance to mention *x*. *12.15*

A large number of grammatical phrases are accepted by both. These include the following:

Mary Morgan *12.16*
Mary Morgan and her sister
some ways of cooking omelettes
the fire hazard
the threat which he made last Saturday
a few rats

But some grammatical phrases are accepted by (12.14) and rejected by (12.15):

I *12.17*
we
?nothing

(The question-mark means that it is debatable whether (12.15) accepts nothing; this may be a matter of dialect.) Some other phrases are accepted by (12.15) but rejected by (12.14):

me *12.18*
us
?how to improve one's tomato crop

There are also a number of frames which accept roughly but not exactly the same phrases as (12.14) or (12.15). For example

He mentioned *x*. *12.19*

accepts very nearly the same phrases as (12.15); but

himself *12.20*

is accepted by (12.19) and rejected by (12.15).

In the cases where a grammatical phrase is accepted by one of the frames above but rejected by another, it usually turns out that the rejection is a fairly gentle one: the ungrammatical sentence is a perturbation of a grammatical one. For example,

*Me will be mentioned afterwards. *12.21*

is a perturbation of

I will be mentioned afterwards. *12.22*

Likewise

?There will be a chance to mention nothing. *12.23*

is (if not grammatical already) a perturbation of

There won't be a chance to mention anything. ***12.24***

Thus we have a cluster of frames which accept or nearly accept the same phrases.

In such cases, we can say that the grammatical phrases which are accepted or nearly accepted by every frame in the cluster form a *phrase-class.*

For example, the frames (12.14), (12.15), (12.19) and similar frames form a cluster which defines the phrase-class of *noun phrases.*

Similarly we can define *nouns* to be those words which are accepted or nearly accepted by frames in the cluster

I saw the *x*. ***12.25***
A *x* will be needed.
(etc.)

Likewise we can define *adjectives* to be those words which are accepted or nearly accepted by frames in the cluster

He produced a *x* book ***12.26***
The *x* things were another matter.
(etc.)

Thus we can use clusters of frames to define classes of words or phrases. Some of the traditional parts of speech can be seen as two or more such classes put together. For instance the class of verbs is got by combining the class of transitive verbs with the class of intransitive verbs, and each of these smaller classes can be defined by a cluster of frames. We need not pursue this here.

Instead of combining classes, we can cut down to smaller classes by eliminating all the phrases which are rejected by some particular frame. For example the phrase-class of *singular noun phrases* consists of the noun phrases which are accepted by

x was mentioned afterwards. ***12.27***

in standard English. Looking through (12.16), we see that the following are singular noun phrases:

Mary Morgan ***12.28***
the fire hazard
the threat which he made last Saturday

while the following are not:

Mary Morgan and her sister *12.29*
some ways of cooking omelettes
a few rats

Actually this is not quite right: on the usual definition, **me**, **him** and **her** are singular noun phrases too, though they are rejected by (12.27). This illustrates the frailty of frame tests.

Why is it that frames give only a rough-and-ready classification of grammatical phrases? Could something better be found?

This is a question of linguistics rather than logic, so only a brief answer is called for here. According to the *transformational syntax* of Noam Chomsky, every sentence has an underlying structure which is different from the written or spoken 'surface structure' of the sentence; in fact there may be several different levels of underlying structure for one sentence. The frame test may give perfect classifications at one of the underlying levels, but these classifications are tangled up as we pass to the surface structure. For example, the words **I** and **me** are not distinguished at the deeper levels; neither are the words **we** and **us**. In this way the main discrepancy between (12.14) and (12.15) disappears at the lower levels. Some features of Chomsky's theory are highly technical – such as his notion that there is a deepest underlying structure, the so-called *deep structure*. But most English speakers would agree that it's perfectly natural to think of **I** and **me** as being fundamentally the same word.

We shall see some more examples of underlying structures in later sections.

One last point. English words have a disarming way of hopping about from one phrase-class to another.

Please **clear** the room. *12.30*
The bell rang the hour loud and **clear**. *12.31*
I didn't get a clear view of him as he ran past. *12.32*

Clear is verb in (12.30), adverb in (12.31) and adjective in (12.32). It will be convenient and natural to adopt the same convention as in our treatment of lexical ambiguity in section 4, and regard **clear** as being three different words which happen to be written alike. **Clear** in (12.30) is a verb, a different word from the adjective **clear** in (12.32).

13. Phrase-markers

Phrase-markers are a handy way of labelling the constituents of a sentence.

We begin with an example. Take the string

The effective dosage varies considerably. ***13.1***

(13.1) forms a grammatical sentence. To express this, we first write

13.2

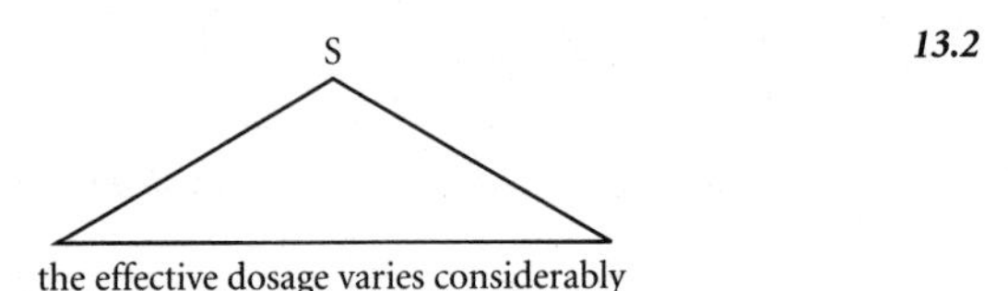

(S means Sentence.) Is there a main break in the sentence? Most people would say that 'the effective dosage' forms a unit, and 'varies considerably' forms another. The first of these units is a noun phrase (=NP); the second is of the type usually called a *verb phrase* (=VP). So we now write

13.3

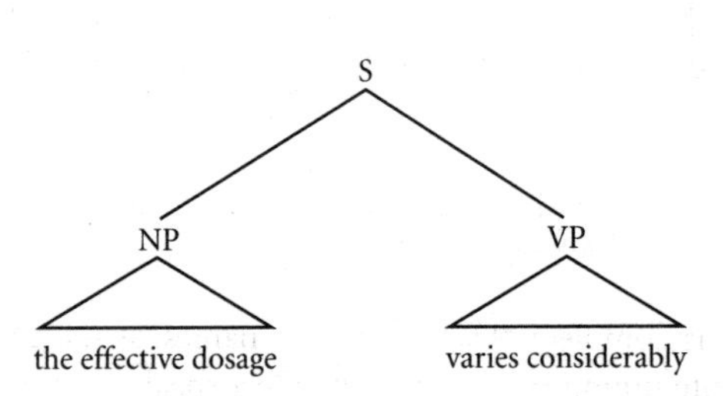

On the right, 'varies' is a verb (=V), and 'considerably' belongs to a phrase-class sometimes called adverbs of manner (=AM). On the left, most people would feel that 'effective dosage' forms a unit. Traditionally one would have called it another noun phrase. But the frame test puts it into the same class as the noun 'dosage', not the same class as 'the effective dosage'. So we call it a noun (=N). 'the' is a determiner (=Det). Thus:

13.4

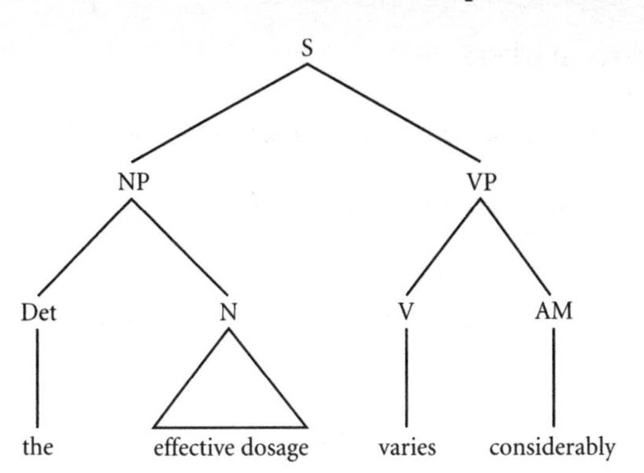

Finally, 'effective' is an adjective (=Adj), and 'dosage' is a noun (=N):

13.5

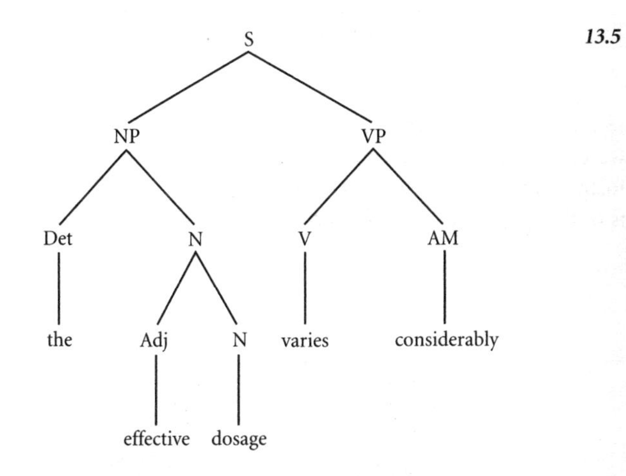

The analysis is complete. Apart from the names of the phrase-classes, which are quite unimportant, probably most speakers of English would make this analysis of (13.1).

(13.5) is an upside-down tree, with the words of the sentence (13.1) strung along the tips of the branches. We call a diagram like (13.5) a *phrase-marker*. The words at the tips of the branches of a phrase-marker are called the *terminal symbols* of the phrase-marker; read from left to right, they form the *terminal string* of the phrase-marker. Thus (13.1) is the terminal string of (13.5). Higher up in a phrase-marker are symbols representing phrase-classes; these symbols are called the *non-terminal symbols* of the phrase-marker. Thus the non-terminal symbols of (13.5) are

S, NP, VP, Det, V, AM, Adj, N. ***13.6***

Wherever a non-terminal symbol appears in a phrase-marker, it tells us that a certain part of the terminal string belongs in the stated phrase-class. Any part of the terminal string which is assigned to a phrase-class in this way is a *constituent* of the terminal string. (This is a refinement of the definition of constituents on page 44.)

For example, we can number the appearances of non-terminal symbols in (13.5) as follows:

13.7

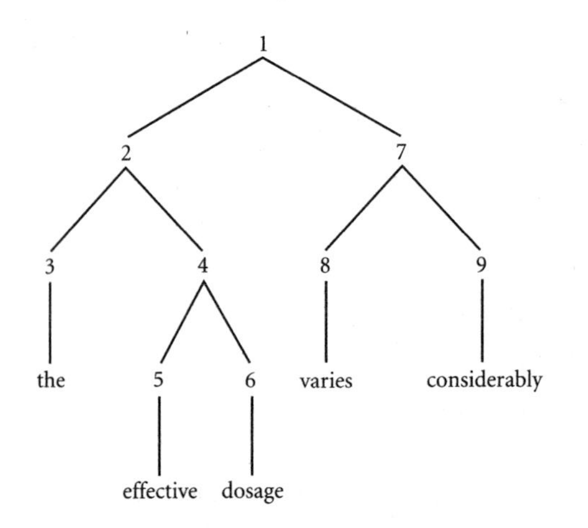

There are nine occurrences of non-terminal symbols, and nine constituents corresponding to these, namely:

13.8

1 [the effective dosage varies considerably]
2 [the effective dosage] varies considerably
3 [the] effective dosage varies considerably
4 the [effective dosage] varies considerably
5 the [effective] dosage varies considerably
6 the effective [dosage] varies considerably
7 the effective dosage [varies considerably]
8 the effective dosage [varies] considerably
9 the effective dosage varies [considerably]

Exercise 13A. Consider the following phrase-marker:

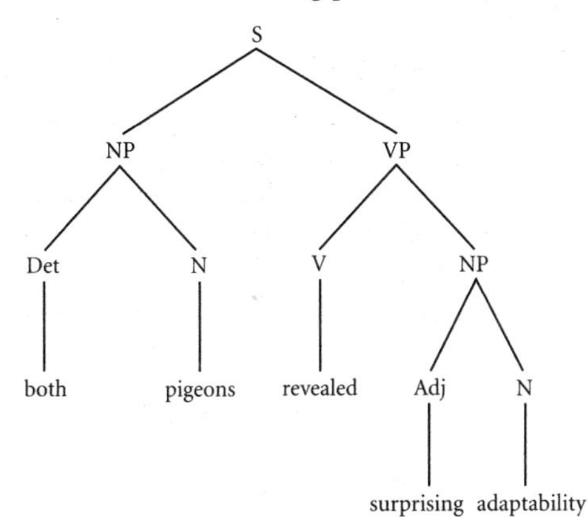

1. What is the terminal string of this phrase-marker?
2. What are the non-terminal symbols of this phrase-marker?
3. List all the constituents of the terminal string.

Exercise 13B. Construct your own phrase-marker for the sentence

The flame melted the wire.

Then list the constituents of this sentence.

So far so good. Probably no speaker of English would object to either of the analyses (13.5) and Exercise 13A. Does this mean that every grammatical English sentence has one correct phrase-marker, which we can find by following our intuitions?

Alas, no. It is very easy to find grammatical English sentences which different people would be inclined to analyse in quite different ways. Often people find they have no feelings at all about how a particular sentence should be divided up. Here is an example of a difficult case:

Liz wants to find her true self. ***13.9***

How should the verb phrase 'wants to find her true self' be split up? One argument is that 'wants to find' is a unit, because it could be rephrased as 'seeks':

Liz seeks her true self. ***13.10***

This would lead to the phrase-marker

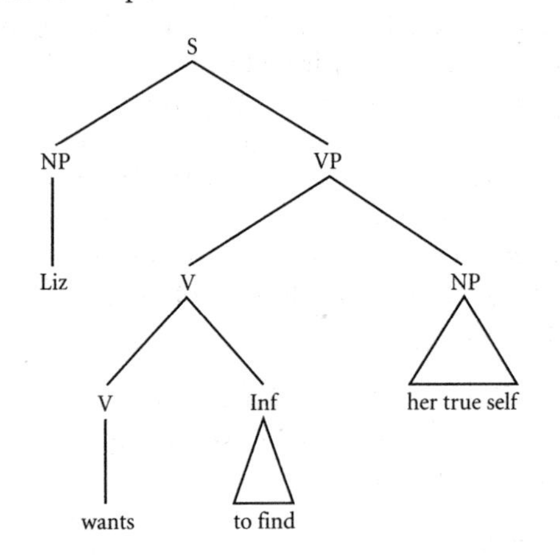

13.11

(Inf = Infinitive; the name is not important.) But another argument says that 'to find her true self' is a unit, because it means 'self-knowledge':

Liz wants self-knowledge. ***13.12***

This argument supports the quite different phrase-marker

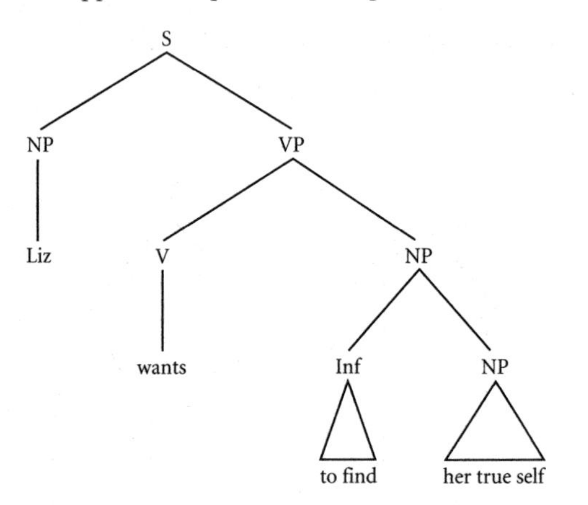

13.13

It is self-deception to think that in cases like (13.9) we can find the right phrase-marker simply by sharpening and training our intuitions. The only way to make progress is to formulate some theory about language, some theory which explains the reasons for the intuitions we do have, and which relates them to the rules of grammar. At this point we touch on current research in linguistics, which is a dangerous field to touch on – the unwary lose their foothold easily.

Nevertheless, one further step into linguistics is worth making. Many grammarians agree with Noam Chomsky that we ought to think of certain grammatical sentences as having been derived from other underlying strings by *transformations*. These transformations may distort the 'true' constituents, by altering the order of the items in the string, or leaving out or repeating or altering parts of the string.

For example, take this sentence:

Bob seized it and broke it up. ***13.14***

We do surely feel that 'broke up' is a unit here, although the word 'it' has squatted in the middle. Compare

Bob broke up the lump. ***13.15***

It's natural to suggest that (13.14) is derived from an underlying string such as

Bob seized it and he broke up it. ***13.16***

This underlying string, which is not a grammatical sentence in English, would presumably have some phrase-marker like

13.17

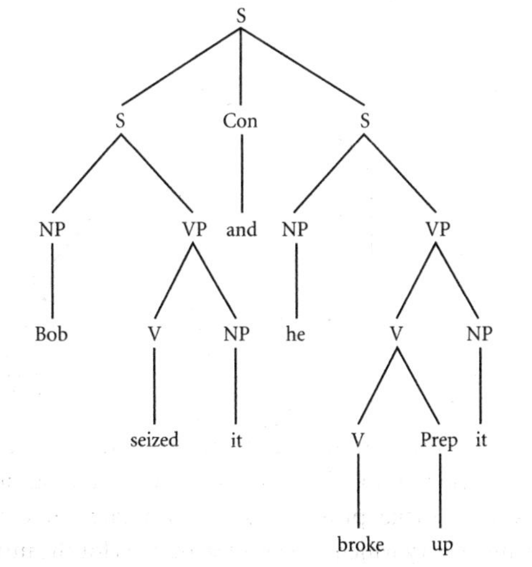

(Con = Conjunction; Prep = Preposition.) (13.17) would then be an *underlying phrase-marker* of the sentence (13.14). 'broke up' is a constituent of (13.16), according to (13.17); we can express this by saying that 'broke up' is an *underlying constituent* of (13.14).

There are plenty of examples of sentences where we would all feel that some underlying phrase-marker would naturally express the 'real' units.

Exercise 13C. Construct underlying phrase-markers to bring out the underlying constituents of the following sentences:

1. I can only see Leila. (i.e., I can see Leila and nobody else.)
2. Did you hear the thunder?
3. I came because I was told to.
4. She will certainly faint.

In some modern theories, the terminal symbols of the underlying phrase-markers need not even be words. For example **I** and **me** might be represented by the same symbol in the underlying phrase-marker (see p. 49). Indeed many contemporary linguists are happy to postulate underlying phrase-markers which are very different from the sentences they are thought to underlie. This will warn us off any deeper probing in this area.

14. Scope

In this section we shall see how phrase-markers can be used for uncovering and curing structural ambiguity (see section 4).

Consider the sentence

The wild animal keeper mopped his brow. ***14.1***

Part of the phrase-marker for (14.1) is quite straightforward:

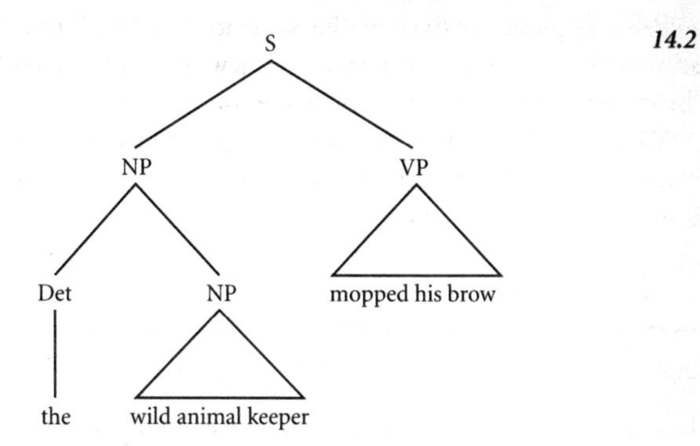

The noun phrase 'wild animal keeper' can be split apart in two ways:

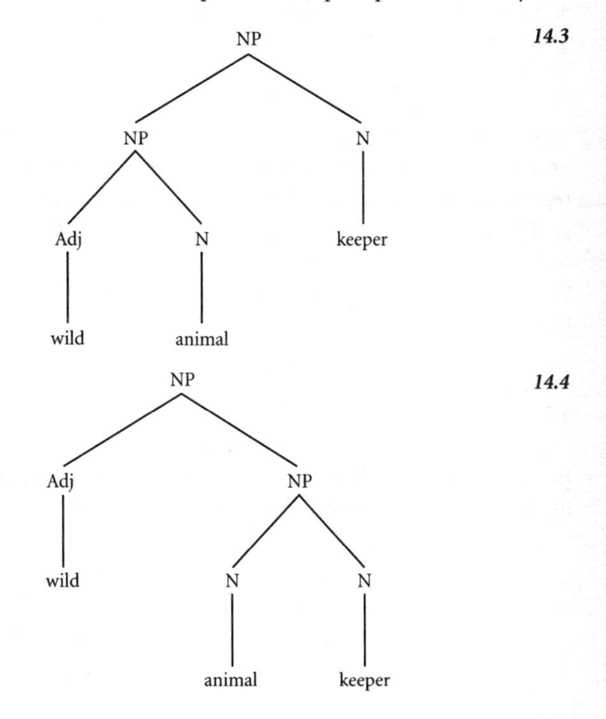

This sentence is totally unlike the problem example (13.9) which we considered earlier. In the present case, the phrases analysed in (14.3) and

(14.4) obviously mean different things. In (14.3), 'wild animal' hangs together, and the whole phrase means

keeper of the wild animals. **_14.5_**

But in (14.4), 'animal keeper' is a unit, and the whole phrase must mean

animal keeper who is wild. **_14.6_**

The two different phrase-markers reveal a structural ambiguity.

The notion of *scope* is helpful here. If S is some sentence with a phrase-marker supplied, and P is a part of S consisting of one or more words, but not the whole of S, then we define the *scope* of P to be the smallest constituent of S which contains both P and something else besides.

For example, if (14.3) is fitted into a phrase-marker for (14.1), then the scope of 'wild' is 'wild animal'; this is a precise way of expressing that 'wild' goes with 'animal'. But if (14.4) is put in place of (14.3), then the scope of 'wild' becomes 'wild animal keeper'; which is a precise way of expressing that 'wild' goes with the whole of 'animal keeper'.

Here is another example, which is important in logic.

The man must be rich or young and good-looking. **_14.7_**

There are two possible phrase-markers (we abbreviate the irrelevant bits):

_14.8_

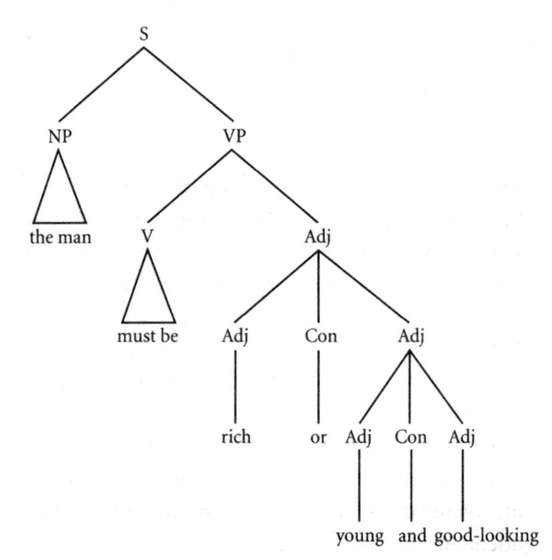

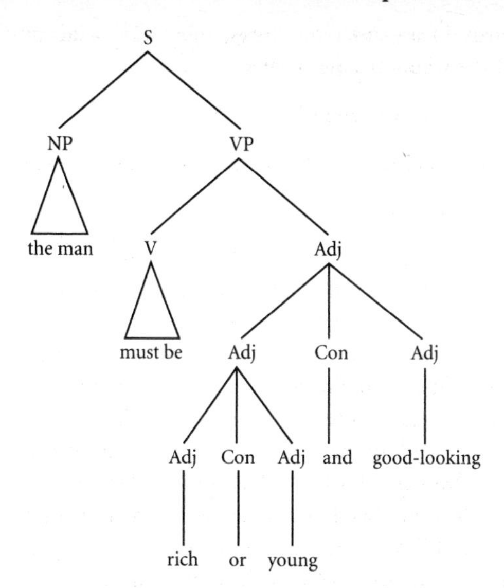

14.9

Clearly the terminal strings of (14.8) and (14.9) mean different things. The man in (14.9) has got to be good-looking, but in (14.8) he can be ugly so long as he is rich. This can be expressed by saying that in (14.9) the scope of 'and' is the whole of

rich or young and good-looking ***14.10***

whereas in (14.8) the scope of 'and' is only

young and good-looking. ***14.11***

In many cases, as we saw in the last section, the 'true' constituents of a sentence can be found only in some underlying phrase-marker. In these cases we must distinguish between the *surface scope* determined by the phrase-marker of the given sentence, and an *underlying scope* determined by an underlying phrase-marker. For example, the surface scope of 'broke' in (13.14) is presumably 'broke it up', whereas according to the underlying phrase-marker (13.17), 'broke' has the underlying scope 'broke up'.

When we write phrase-markers in order to clear up structural ambiguities, it is usually unnecessary to present a whole phrase-marker; an abbreviated one using the triangle notation (as (14.8) and (14.9)) is usually quite adequate.

Exercise 14. Write abbreviated phrase-markers (underlying if necessary) to explain the structural ambiguities in the following sentences:

1. The ambassador was ordered to leave in the morning.
2. Amos shaved and played with the cat.
3. I won't vote as a protest.
4. He only relaxes on Sundays.

15. Context-free Grammars†

A grammar for a language is basically a set of rules which tell us how to construct the grammatical sentences of the language. The rules should produce all the grammatical sentences of the language, and they should not produce anything else. It is reasonable to expect a grammar to tell us what the constituents of a sentence are as well.

The simplest grammars which are any use at all are the so-called *context-free grammars*, or *CF grammars* for short. We shall define them in a moment. They are hopelessly inadequate to deal with a full-grown language like English. But they are competent to handle the basic formal languages of logic, as we shall see later.

The main notion in CF grammars is that of a *CF rewriting rule*. By a *CF rewriting rule* we mean an expression of the form

$$A \Rightarrow B_1 \dots B_n \qquad \textit{15.1}$$

where A, $B_1, \dots, B_n$ are written expressions, called the *symbols* of this rule. Exactly one symbol must appear on the left of '⇒', while one or more symbols can appear on the right of '⇒'. (15.1) should be thought of as standing for an upside-down tree

15.2

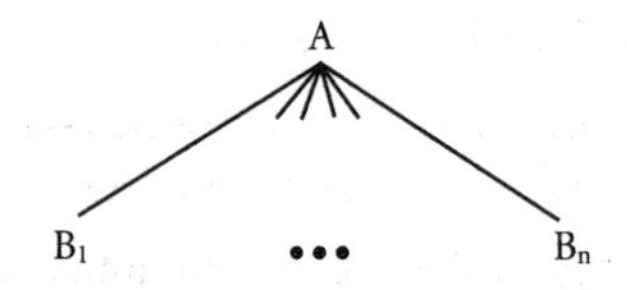

A CF grammar is defined to be a list of CF rewriting rules.

† This section is not used later, except in mathematical sections.

For example, here is a CF grammar:

1 S ⇒ NP VP **_15.3_**
2 NP ⇒ Det N
3 N ⇒ Adj N
4 VP ⇒ V NP
5 Det ⇒ the
6 Det ⇒ this
7 Adj ⇒ old
8 N ⇒ fool
9 N ⇒ sentence
10 V ⇒ ignored

The numbers on the left in (15.3) are not part of the rules; they simply mark the place of each rule in the list.

Rule number 1 of (15.3) could also be written in the form

_15.4_

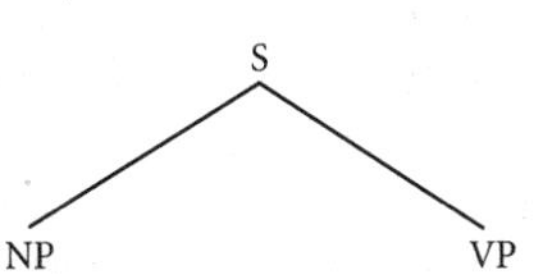

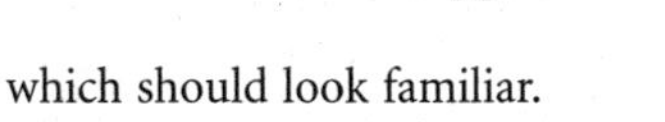

which should look familiar.

If ***C*** is a CF grammar, then we refer to the left-hand symbol of the first rule in ***C*** as the *initial symbol* of ***C***. If a symbol occurs on the left-hand side of some rule in ***C***, then we say the symbol is a *non-terminal symbol* of ***C***; symbols which occur only on the right-hand side of rules in ***C*** are called *terminal symbols* of ***C***.

For example, the initial symbol of the CF grammar (15.3) is '**S**', and its non-terminal symbols are

S, NP, N, VP, Det, Adj, V. **_15.5_**

The terminal symbols of this CF grammar are

the, this, old, fool, sentence, ignored. **_15.6_**

If ***C*** is a CF grammar and ***P*** is a phrase-marker, then we say that ***C*** generates ***P*** if the following three things are true:

a. the symbol at the top of ***P*** is the initial symbol of ***C***;
b. every terminal symbol of ***P*** is a terminal symbol of ***C***;
c. every step between the top and the bottom of ***P*** is one of the CF rewriting rules of ***C***.

We say that ***C** generates* a string if it generates a phrase-marker with the string in question as its terminal string.

For example, the CF grammar (15.3) generates the following phrase-marker:

15.7

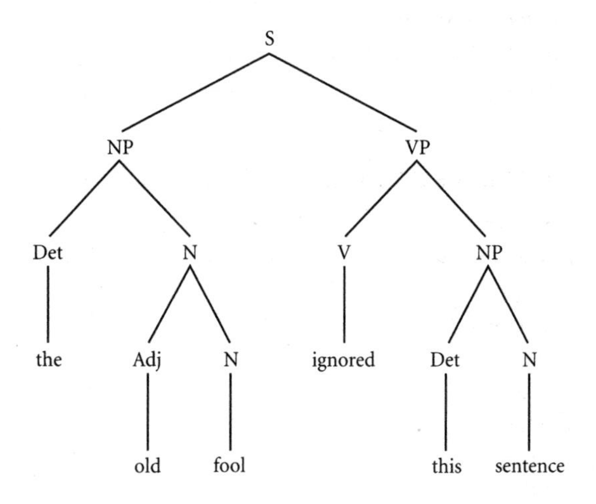

It follows that (15.3) generates the terminal string of (15.7), which is

The old fool ignored this sentence. ***15.8***

The CF grammar (15.3) also generates the phrase-marker

15.9

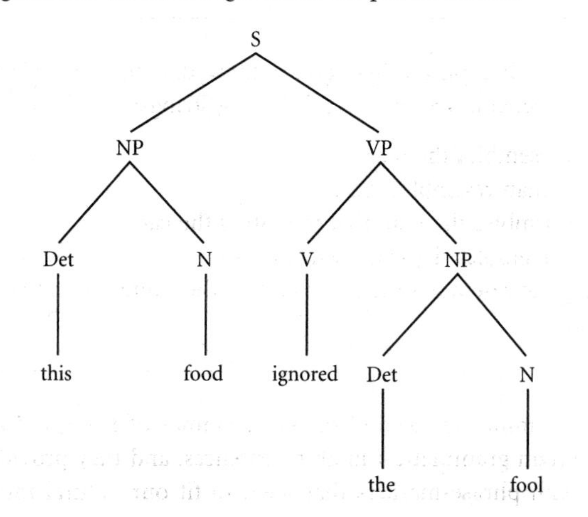

and so this grammar also generates the sentence

This fool ignored the fool. ***15.10***

The same CF grammar also generates

This fool ignored this fool. ***15.11***
The old old sentence ignored the fool.
The sentence ignored the sentence. (etc.)

Exercise 15A. Here is a CF grammar:

1 S ⇒ NP VP
2 VP ⇒ NP
3 NP ⇒ Det N
4 N ⇒ N that VP
5 Det ⇒ the
6 N ⇒ Adj N
7 Adj ⇒ white
8 N ⇒ man
9 N ⇒ cat
10 V ⇒ resembles

What are the initial symbol, the non-terminal symbols, and the terminal symbols of this CF grammar?

Exercise 15B. Give phrase-markers to show that the CF grammar of Exercise 15A generates each of the following strings:

1. the man resembles the man
2. the white man resembles the cat
3. the cat resembles the man that resembles the cat
4. the man resembles the white white cat
5. the man that resembles the cat resembles the white cat that resembles the man

---

The CF grammar (15.3) and the CF grammar of Exercise 15A each generate certain grammatical English sentences, and they provide these sentences with phrase-markers that seem to fit our natural intuitions.

These two CF grammars are therefore reasonable grammars for two fragments of English. (They might well seem less reasonable if we tried to extend them to cover larger classes of grammatical sentences.)

It's interesting to note that both these two CF grammars generate infinitely many different strings. In (15.3), rule 3 can be applied over and over again, as many times as we like; each time we apply it, the effect is to add another 'old'. Thus the grammar generates sentences such as

The old old old old old fool ignored this sentence. ***15.12***

In the CF grammar of Exercise 15A, rule 4 has a similar effect.

Exercise 15C. A one-year-old child is overheard making the following remarks:

boy. sock. mommy. allgone boy. allgone mommy. byebye sock. byebye mommy. boy off. sock off. sock on. mommy on. mommy fall.

Construct an appropriate CF grammar for the language of this child. (Remember that these are only a sample of what the child can produce; you have to work out what the implicit rules are.)

Logical Analysis

In the next four sections we encounter one of the chief skills of logic: that of logical analysis. We first select a small number of ways of combining short sentences into longer sentences. Then we show that very many sentences, if they are not already built up in these ways, mean the same as certain other sentences which are so built up. *Logical analysis* consists in finding these other sentences; it stands somewhere between translating and paraphrasing.

Though logicians agree about how to analyse, they disagree about the purpose. Some regard themselves as uncovering the 'real form' of the sentence they analyse, while others see logical analysis as part of an enterprise to replace English by a new and more rational language. We need not enter the controversy; for us the justification is that a tableau for logically analysed sentences can be constructed quite mechanically, according to strict mathematical rules.

16. Sentence-functors and Truth-functors

Some of the sentences in the examples of section 10 were particularly easy to handle, because they had shorter constituents which were sentences. For example,

[Cobalt is present] but [only a green colour appears]. ***16.1***

(The constituent sentences are marked with square brackets.) We can analyse such a sentence into its constituent sentences

Cobalt is present. ***16.2***
Only a green colour appears.

together with the matrix which contained them:

ϕ but ψ. ***16.3***

The Greek letters 'ϕ' and 'ψ' are called *sentence variables*, which means that they are symbols standing for sentences. Here they mark the holes where the constituent sentences should go. We shall also use the Greek letter 'χ' as a sentence variable; we may add subscripts too, as in 'ϕ_1', 'ϕ_2', etc.†

There are several different senses in which a symbol can be used to 'stand for' expressions. Rather than catalogue these senses, we shall introduce them as and when they are needed; the context should always make clear what is meant. Where a variable serves to mark a hole that can be filled with an expression of a certain sort, the variable is said to have a *free* occurrence; in (16.3) both the occurrences of variables are free.

The matrix (16.3) is an example of a *sentence-functor.* More precisely, a *sentence-functor* is defined to be a string of English words and sentence variables, such that if the sentence variables are replaced by declarative sentences, then the whole becomes a declarative sentence with the inserted sentences as constituents. Here is a selection of sentence-functors:

It's a lie that ϕ. ***16.4***
Many authorities have noted that ψ. ***16.5***
She went and bought some fish, then ϕ. ***16.6***
If ϕ, then ψ. ***16.7***
ϕ because ψ, unless χ. ***16.8***
Since he swears that ψ, we can take it that ψ. ***16.9***

Every occurrence of a sentence variable in a sentence-functor is free.

Sentence-functors are classified by the number of different sentence variables they contain. The three examples (16.4)–(16.6) each contain just one sentence variable, so that they are described as *1-place* sentence-functors. (16.3) and (16.7) are *2-place*, while (16.8) is *3-place.*

Note that (16.9) is *1-place*, since only one sentence variable occurs in it, even though it occurs twice. When a sentence variable is repeated in a sentence-functor, this is understood to mean that the sentence variable must be replaced by *the same sentence at each occurrence.* For example the holes in (16.9) can be filled to form the sentence

† ϕ pronounced *fie*; χ pronounced *khi*; ψ pronounced *psi*; all rhyming with *sky.*

Since he swears that he was at home, we can take it that he was at home. ***16.10***

They cannot be filled to produce

Since he swears that he was at home, we can take it that he is not guilty. ***16.11***

Exercise 16A. Analyse the following sentence into one 4-place sentence-functor (with sentence variables 'ϕ_1', 'ϕ_2', 'ϕ_3' and 'ϕ_4') and four constituent sentences:

I scattered the strong warriors of Hadadezer, and then at once I pushed the remnants of his troops into the Orontes, so that they dispersed to save their lives; Hadadezer himself perished.

Returning to example (16.1), we see that this sentence is true precisely when both the constituent sentences (16.2) are true. In fact if we replace 'ϕ' and 'ψ' in (16.3) by declarative sentences, then the whole resulting sentence will be true precisely if both the added sentences are true. We can express this in a chart, as follows:

ϕ	ψ	ϕ but ψ.
T	T	T
T	F	F
F	T	F
F	F	F

16.12

Here T = True and F = False; thus the third row of the table (16.12) indicates that if in 'ϕ but ψ' we replace 'ϕ' by a false sentence and 'ψ' by a true one, then the whole resulting sentence is false. This chart (16.12) is called a *truth-table* for the sentence-functor (16.3).

Likewise we can write down a truth-table for the sentence-functor

It's true that ϕ. ***16.13***

as follows:

ϕ	It's true that ϕ.
T	T
F	F

16.14

Similarly the sentence-functor

Either it's true that ϕ or it's not true that ϕ. *16.15*

has a truth-table. The sentence variable 'ϕ' must be replaced by the same declarative sentence at both places, so as to yield sentences like

Either it's true that high taxes are inflationary or it's not true that high taxes are inflationary. *16.16*

High taxes may be inflationary, or they may not be; (16.16) is equally true in either case. The truth-table is accordingly:

ϕ	Either it's true that ϕ or it's not true that ϕ.
T	T
F	T

16.17

However, not every sentence-functor has a truth-table. Sometimes the truth-values of the constituent sentences are not enough by themselves to determine the truth-value of the whole. For example, the sentence-functor

I know that ϕ. *16.18*

has only the partial truth-table:

ϕ	I know that ϕ.
T	–
F	F

16.19

Nobody can know something that is false; hence the second row of (16.19) shows Falsehood. But there are truths which I know and truths which I don't know, and this compels us to leave the first row blank. For the sentence-functor

It is often asserted that ϕ. *16.20*

the partial truth-table is even more sparse:

ϕ	It is often asserted that ϕ.
T	–
F	–

16.21

A sentence-functor which has a truth-table is called a *truth-functor*; thus (16.3), (16.13) and (16.15) are truth-functors, while (16.18) and (16.20) are not. The next few sections will be entirely concerned with truth-functors. Sentence-functors that are not truth-functors are much harder to handle; we shall consider some examples in section 42.

Exercise 16B. Write out a truth-table or a partial truth-table (as appropriate) for each of the following sentence-functors:

1. It's a lie that ϕ.
2. ϕ because ψ.
3. ϕ whenever ψ.
4. If ϕ, then ϕ.
5. Whether or not ϕ, what will be will be.
6. Whether or not ϕ, smoking causes cancer.

To make a logician happy, find him some constituent sentences. Very often a sentence will yield us more constituent sentences if we allow ourselves to *paraphrase* it, i.e., to replace it by another sentence which means the same thing. If two declarative sentences mean the same thing, then they have the same truth-value in all situations; so there is no harm in replacing one by the other in an argument which is being tested for validity, or in a set of sentences which is being tested for consistency.

To illustrate the powers of paraphrase, consider the following morsel of seventeenth-century legal prose:

> If the process be legal, and in a right Court, yet I conceive that His Majesty alone, without assistance of the Judges of the Court, cannot give judgment. *16.22*

The following paraphrase of (16.22) contains four constituent short sentences, which we mark with brackets:

> I conceive that, if [the process is legal] and [the process is in a right court], yet if [the Judges of the Court are not assisting His Majesty], [His Majesty cannot give judgment]. ***16.23***

The sentence (16.23) can be got by filling the gaps in the sentence-functor

> I conceive that, if ϕ_1 and ϕ_2, yet if ϕ_3, ϕ_4. ***16.24***

Many sentences can be paraphrased in a simple way to elicit short constituent sentences. (In quite a few cases, linguists would regard such a paraphrase as bringing to the surface an underlying constituent.)

Exercise 16C. Paraphrase the following so as to find constituent sentences:

1. I am aware of your intention to sue.
2. He regrets not having married Suzy.
3. He completed his task before the end of the week.
4. An accident to the train resulted from the failure of its brakes.
5. Your Majesty may be pleased to take notice of the great mischiefs which may fall upon this kingdom if the intentions which have been credibly reported, of bringing in Irish and foreign forces, shall take effect.

---

17. Some Basic Truth-functors

In this section we shall introduce the five truth-functors most commonly used in logic; each of them has a special symbol to represent it. At the same time we shall mention some English expressions which can be paraphrased by means of these truth-functors.

(i) The *negation* truth-functor 'It is not true that ϕ.'

In symbols, this truth-functor is written '$\neg\phi$'. It yields true sentences precisely when false sentences are put for 'ϕ', so that its truth-table is

ϕ	$\neg\phi$
T	F
F	T

17.1

'$\neg\phi$' is called the *negation* of the sentence ϕ.† '$\neg$' is pronounced 'not'.

Here are some other ways in which English expresses the sense of '$\neg$':

I am **not** a Dutchman. *17.2*
$\neg$I am a Dutchman.

She **didn't** say anything. *17.3*
$\neg$she said something.

George Washington **never** told a lie. *17.4*
$\neg$George Washington sometimes told a lie.

I **hardly** think he will reach Athens in that old bus. *17.5*
$\neg$I think he will reach Athens in that old bus.

It **isn't as if** he needs the money. *17.6*
$\neg$he needs the money.

None of these paraphrases is perfect. For example, the first sentence of (17.3) implies that there is some woman under discussion – for otherwise we have referential failure. But 'She said something' is false if there is no woman to be referred to by 'she', so that in such a situation the second sentence of (17.3) is true.

Similarly if there is no old bus in the situation, then the first sentence of (17.5) is false while the second is true.

In fact, when one adds words such as **not** or **never** to an English sentence, this cancels some of the implications of the sentence, but it usually leaves other implications intact. A famous example is

I have not stopped beating my wife. *17.7*

By contrast the symbol '$\neg$' cancels all the implications of the original sentence.

Since they are not completely accurate, paraphrases like (17.2)–(17.6) may lead to mistakes in logic. In practice this happens very rarely, and the

† In this sentence, the symbol 'ϕ' is being used to talk about sentences, not to mark a hole where a sentence can be put. Thus the two occurrences of 'ϕ' here are not free.

experienced logician knows when to take care. (We shall return to this in section 28.)

(ii) The *conjunction* truth-functor 'ϕ and ψ'.

This truth-functor is written '$[\phi \wedge \psi]$'. It yields a false sentence unless truths are put for both 'ϕ' and 'ψ', in which case it yields a truth. (Comparing with (7.1), one can see that we are assuming the weak reading of **and**.) Hence the truth-table is

ϕ	ψ	$[\phi \wedge \psi]$
T	T	T
T	F	F
F	T	F
F	F	F

17.8

'$[\phi \wedge \psi]$' is called the *conjunction* of the sentences ϕ and ψ; ϕ and ψ are its *conjuncts*. It is pronounced 'ϕ and ψ'.

Here are some other ways in which English expresses the sense of this truth-functor:

Although it was raining, he ran out in his vest. *17.9*
[it was raining ∧ he ran out in his vest]

The powder contains sulfur **and** magnesium. *17.10*
[the powder contains sulfur ∧ the powder contains magnesium]

The method is simple **but** effective. *17.11*
[the method is simple ∧ the method is effective]

Neither I nor my wife speak German. *17.12*
[I don't speak German ∧ my wife doesn't speak German]

The damping is effected by the water roller, **which** can be found above the plate cylinder. *17.13*
[the damping is effected by the water roller ∧ the water roller can be found above the plate cylinder]

Notice that in (17.10) **and** is between nouns instead of sentences. The word **and** can occur between adjectives too, and other parts of speech. We can usually rephrase the whole sentence so that **and** occurs just between sentences, allowing a translation by '∧':

Your review of my book was **both** insulting **and** inaccurate. ***17.14***

[your review of my book was insulting $\wedge$ your review of my book was inaccurate]

He coughs often **and** loudly. ***17.15***

[he coughs often $\wedge$ he coughs loudly]

But sometimes this kind of rephrasing is quite wrong:

New York **and** Cairo are over a thousand miles apart. ***17.16***

NOT: [New York is over a thousand miles apart $\wedge$ Cairo is over a thousand miles apart]

My sister wants a black **and** white cat. ***17.17***

NOT: [my sister wants a black cat $\wedge$ my sister wants a white cat]

Twenty people were rounded up **and** shot. ***17.18***

NOT: [twenty people were rounded up $\wedge$ twenty people were shot]

(iii) The *disjunction* truth-functor 'Either ϕ or ψ, or both'. This truth-functor is written '$[\phi \vee \psi]$'; its truth-table is

17.19

ϕ	ψ	$[\phi \vee \psi]$
T	T	T
T	F	T
F	T	T
F	F	F

The table shows that '$[\phi \vee \psi]$' yields a truth in every case but one; the one case is where 'ϕ' and 'ψ' are both replaced by false sentences. '$[\phi \vee \psi]$' is called the *disjunction* of the sentences ϕ and ψ, and ϕ and ψ are called its *disjuncts*. It is pronounced 'ϕ or ψ'.

Here are some ways in which English expresses the sense of this truth-functor:

There will be a stiff wages policy, **or** we shall see massive unemployment. ***17.20***

[there will be a stiff wages policy $\vee$ we shall see massive unemployment]

He is a fool **or** a liar. *17.21*
[he is a fool $\vee$ he is a liar]

We'll go to the seaside **unless** it rains. *17.22*
[we'll go to the seaside $\vee$ it will rain]

We take (17.20) to be true (but possibly misleading) if there will be *both* a stiff wages policy *and* massive unemployment – this is the top line of the truth-table (17.19). Some people would regard it as false in this situation. They take the strong reading, while we take the weak one (see section 7).

In (17.22), note that the underlying constituent is 'it will rain', not 'it rains'. After **unless**, English uses present tense instead of future; we have to remember to put the verb back into future form when we introduce '$\vee$'.

There are a few unusual sentences in which **or** means something more like **and**, and it should be translated to '$\wedge$':

Uri Geller can read your mind, **or** he can bend your spoons. *17.23*
[Uri Geller can read your mind $\wedge$ he can bend your spoons]

(iv) The *arrow* truth-functor '$[\phi \rightarrow \psi]$' (sometimes called *material implication*).

There is no neat and exact way of expressing this truth-functor in ordinary English, though several English expressions come close to it. The best way to define it is by its truth-table, which is

17.24

ϕ	ψ	$[\phi \rightarrow \psi]$
T	T	T
T	F	F
F	T	T
F	F	T

As the table shows, the only way to make a false sentence out of '$[\phi \rightarrow \psi]$' is to put a true sentence for 'ϕ' and a false sentence for 'ψ'. '$\rightarrow$' is best read as 'arrow'.

Here are some English phrases which can be expressed with this truth-functor:

If the paper turns red, **then** the solution is acid. *17.25*
[the paper will turn red $\rightarrow$ the solution is acid]

The paper will **only** turn red **if** the solution is acid. *17.26*
(or: The paper will turn red **only if** the solution is acid.)
[the paper will turn red → the solution is acid]

You will get a room **provided** you have no pets. *17.27*
[you have no pets → you will get a room]

Assuming that the timer is correctly set, the relay will close after two minutes. *17.28*
[the timer is correctly set → the relay will close after two minutes]

If there are any more patients, I shall be home late. *17.29*
[there are some more patients → I shall be home late]
NOT: [there are any more patients → I shall be home late]

The first sentence of (17.25) excludes just one possible state of affairs, namely that the paper will turn red and the solution is not acid. As we see from the table (17.24), this is precisely the case which is ruled out by the second sentence of (17.25) too. We can test the accuracy of the other translations (17.26)–(17.29) in the same way. Note the dramatic effect of adding **only** in (17.26) compared with (17.25) – it shifts the **if** to the other clause.

In some of these examples, we have to seek out the underlying constituents. Thus in (17.25) the underlying constituent sentence is 'the paper will turn red'; English drops the future tense after **if**. In (17.29) we have to paraphrase to remove **any**.

Some of these translations may raise doubts. For example, surely the first sentence of (17.25) implies there is some kind of connection between the redness and the acidity? And surely it suggests that if the paper does not turn red, then the solution is not acid? Neither of these things is conveyed by our translation. We take the view that although somebody might well assume these things if she heard the first sentence of (17.25), they are not actually stated in that sentence. In the terminology of section 7, we adopt the weak reading.

There are some cases where **if** should definitely not be translated by '→'. Here are three examples; more are given in section 18.

The choir was sensitive, **if** a little strained. *17.30*
[the choir was sensitive ∧ the choir was a little strained]

> **If** you want to wash your hands, the bathroom is first on the left. *17.31*

> I won't sing, **even if** you pay me £1000. *17.32*

In fact (17.32) is true in just the same situations as

> I won't sing. *17.33*

The difference lies in the emphasis alone.

(v) The *biconditional* truth-functor 'ϕ if and only if ψ'.

This truth-functor is written '$[\phi\leftrightarrow\psi]$'. Its truth-table is

17.34

ϕ	ψ	$[\phi\leftrightarrow\psi]$
T	T	T
T	F	F
F	T	F
F	F	T

Note the third line, which distinguishes '$\leftrightarrow$' from '$\rightarrow$'. '$\leftrightarrow$' is pronounced 'if and only if'.

Two other phrases which carry the sense of '$\leftrightarrow$' are **precisely if** and **just if**:

> The number is even **precisely if** it's divisible by two. *17.35*
> [the number is even $\leftrightarrow$ it's divisible by two]

> The company has to be registered **just if** its annual turnover is above £15,000 *17.36*
> [the company has to be registered $\leftrightarrow$ its annual turnover is above £15,000]

In America, but not normally in Britain, **just in case** is used in the same way.

This completes our list of basic truth-functors. There is one important point to bear in mind before we face any exercises. Our object is to analyse complex sentences into shorter ones, so that we can extract the shorter sentences in a tableau. Now if a complex sentence contains cross-referencing, the references of some of its parts may change when we extract the constituent sentences; we may even face referential failure. For example, in the first sentence of (17.35), the pronoun 'it' refers back to the number mentioned at the beginning. But if we take the second constituent sentence on its own:

It's divisible by two. ***17.37***

there's nothing to determine what 'it' refers to. To avoid this, you should always *try to eliminate cross-referencing* when you translate. In the example just given, you should replace the translation in (17.35) by

[the number is even $\leftrightarrow$ the number is divisible by two] ***17.38***

The same applies to (17.23) and (17.36). There are some more examples in the exercises below.

Exercise 17. Express each of the following sentences as faithfully as possible, using the truth-functors introduced in this section; remove cross-referencing where possible.

1. No dogs will be admitted.
2. The brain is bisected, yet the character remains intact.
3. Unless the safety conditions are tightened, there is going to be a nasty accident.
4. Supposing you're right, I stand to lose a lot of money.
5. You broke the law if and only if the agreement formed a contract.
6. If anybody calls, I shall pretend I am designing St Paul's.
7. Schubert is terrific, and so is Elvis Costello.
8. This is Bert Bogg, who taught me that limerick I was quoting yesterday.
9. You can only claim the allowance if you earn less than £160 a week.
10. Liszt is horrible, and the same goes for Vivaldi.
11. She needs all the help she can get, being a single parent.
12. The elder son was highly intelligent, while the younger had learning difficulties.
13. Her performance lacked zest.
14. If he gets anything right at all, he'll pass.
15. Either the metal will stretch, or it will snap.

---

18. Special Problems with '→' and '∧'

The truth-functor symbols '→' and '∧' raise some special problems, which deserve a few words of warning. The points at issue are fairly subtle; this section can be left out without breaking the continuity.

We begin with '→'.

In a sentence of the form 'If A then B', we often find cross-referencing from B to A, and it may be impossible to paraphrase so as to remove the cross-referencing. Here's a fairly mild example:

> If the tonsils are removed, the adenoids are often cut out too. ***18.1***

If we ask 'Which adenoids?', the answer is: the adenoids of whoever has his or her tonsils cut out. But thousands of people have their tonsils cut out, so that there is no question of finding a phrase which pins us down to just one set of adenoids. Even if we could find one, we should hardly interpret (18.1) as saying that one particular person's adenoids are often cut out! So we must not analyse (18.1) as

> [the tonsils are removed → the adenoids are often cut out]. ***18.2***

For similar reasons we must avoid such translations as the following:

> If you poured in the sulfuric acid, the solution would turn muddy. ***18.3***
> NOT: [you poured in the sulfuric acid → the solution would turn muddy]

> If you had poured in the sulfuric acid, the solution would have turned muddy. ***18.4***
> NOT: [you had poured in the sulfuric acid → the solution would have turned muddy]

The English sentences in (18.3) and (18.4) are examples of *subjunctive conditionals*; they say what would happen in hypothetical states of affairs. The second clause refers to the hypothetical state of affairs described by the first clause, so that there is a cross-reference.

Exercise 18A. Which of these sentences (with apologies to Dr Spock) can be translated by means of '→' without cross-referencing?

1. If the nappies are becoming hard, you can soften them by using a water conditioner.
2. If it contains soap, this helps in removing stains.
3. If an injured child has not already built his own protection from toxoid inoculations, it is sometimes hard to decide whether horse serum is necessary.
4. If most of his former protection has worn off, his new vaccination develops much like the previous one.
5. If a vaccination doesn't take, it doesn't mean that the person is immune.
6. If your baby is colicky, he may be soothed when you first pick him up.

We turn to '$\wedge$'.

The symbol '$\wedge$' can sometimes be used to eliminate the relative pronouns **which** and **who**. For example,

> The policeman, who was watching through binoculars, ducked just in time. ***18.5***

can be analysed as

> [the policeman was watching through binoculars $\wedge$ the policeman ducked just in time] ***18.6***

See (17.13) for another example.

However, there is another sentence very like (18.5), which must not be analysed in this way. The sentence is

> The policeman who was watching through binoculars ducked just in time. ***18.7***

(Note the commas.) As it occurs in (18.7), the phrase 'who was watching through binoculars' serves to indicate which of several policeman is being talked about; in this use it is said to be *restrictive*. The same phrase in (18.5) serves, not to pick out one policeman from several, but to say something about a policeman who has already been picked out; we say it is *non-restrictive* in (18.5). The truth-functor symbol '$\wedge$' *must not be used to replace* **which** *or* **who** *in restrictive phrases.*

We can make the same distinction with adjectives. When **clever** is used non-restrictively, we may be able to split it off with the help of '∧':

My clever husband has found a tax loophole that saves us £5000. ***18.8***
[my husband is clever ∧ my husband has found a tax loophole that saves us £5000].

Karel is a clever boy. ***18.9***
[Karel is clever ∧ Karel is a boy]

When it is used restrictively, no such analysis is possible:

The clever twin was always teasing her dim-witted sister. ***18.10***
NOT: [the twin was clever ∧ the twin was always teasing her dim-witted sister]

Even when we find an adjective in a non-restrictive posture, we may be unable to split it off with '∧', because the meaning requires it to stay attached to a particular noun. For example:

Arturo is a **famous pianist.** ***18.11***
NOT: [Arturo is famous ∧ Arturo is a pianist]

(The second sentence is true if, for example, the pianist Arturo is famous only as a female impersonator.)

You are a **perfect stranger.** ***18.12***
NOT: [you are perfect ∧ you are a stranger]

The company car was a **small compensation.** ***18.13***
NOT: [the company car was small ∧ company car was a compensation]

Amaryllis is **my daughter.** ***18.14***
NOT: [Amaryllis is mine ∧ Amaryllis is a daughter]

Exercise 18B. Which of these sentences can be analysed by means of '∧'?

1. Stavros is a so-called radical.
2. Stavros is a former radical.
3. The thoroughly pleasant evening concluded with a waltz.

4. Their next encounter was more restrained.
5. Her aunt, who from her earliest youth
   Had kept a strict regard for Truth,
   Attempted to believe Matilda.
6. The animal that you saw was probably a fox.
7. Marianne is a teacher, who should have known better.
8. Britain, once a superpower, is now seeking a new role.
9. I can see Don growing into a bespectacled pedant.

Some arbiters of English style recommend using **which** or **who** in non-restrictive phrases and **that** in restrictive ones. Others say that one should use **that** only for inanimate objects. English has never followed either of these rules, either in conversation or in high written style. There seem to be just two safe generalizations: in written English the comma rule that distinguished 18.5 from 18.7 is well established, and in written and spoken English the use of **that** in non-restrictive phrases is much less common today than it was in Shakespeare's time. Beyond these you must rely on your common sense – as always in logical analysis.

19. Analysis of Complex Sentences

We wish to rewrite the following sentence using truth-functors:

> This female bearded reedling has no black marks underneath, and its head is tawny. ***19.1***

There are an **and** and a **no** to contend with. Two translations suggest themselves:

> [¬this female bearded reedling has black marks underneath ∧ this female bearded reedling has a tawny head] ***19.2***

> ¬[this female bearded reedling has black marks underneath ∧ this female bearded reedling has a tawny head] ***19.3***

(19.2) is right and (19.3) is wrong. Why?

The answer is a matter of scope. In (19.1) the scope of 'and' is the whole sentence:

19.4

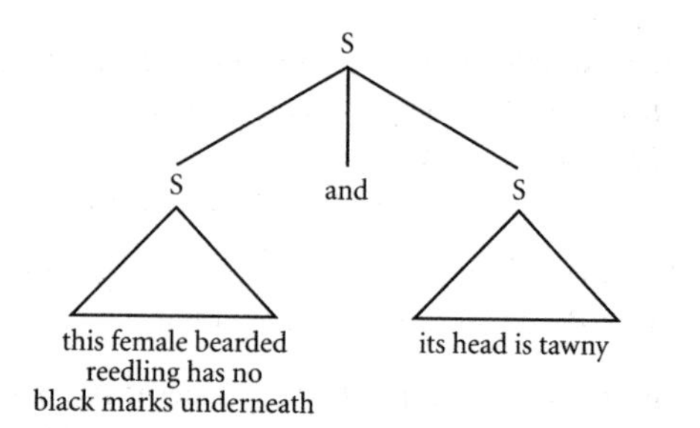

We shall express this by saying that the *overall form* of (19.1) is 'ϕ and ψ'. The correct translation (19.2) likewise has the overall form '$[\phi \wedge \psi]$':

19.5

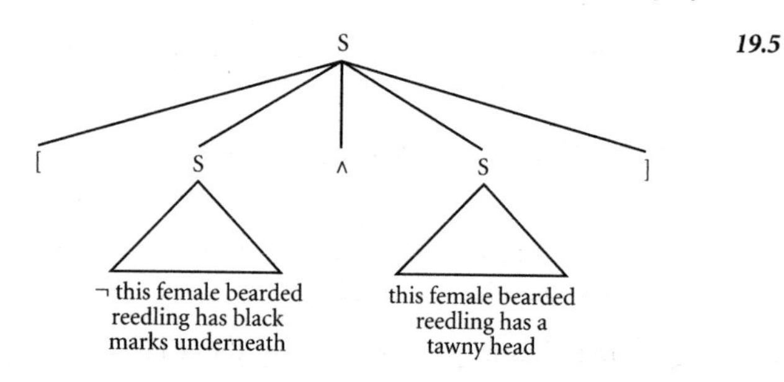

But in the incorrect translation, the scope of '$\wedge$' is only part of the sentence, and the overall form is '$\neg\,\phi$':

19.6

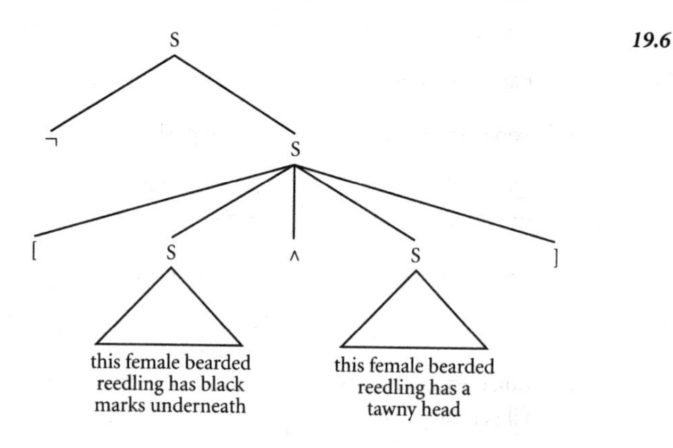

This example shows why we included the two brackets in our notation; without them it would be impossible to tell whether the scope of '$\wedge$' included '$\neg$' or not.

To translate a complex sentence with more than one truth-functor, *start with the truth-functor of largest scope, as shown by the overall form of the sentence; then work inwards.*

The following example will show how. We shall translate:

> If the battery is flat, then the starter will be dead, and you won't get the car started unless we push it. ***19.7***

We first look for the overall form of (19.7). Does it say 'If ϕ then ψ' or 'ϕ and ψ', or 'It's not true that ϕ', or what? On the natural reading, the part after 'then' forms a unit, so the answer is 'If ϕ then ψ'. The truth-functor of largest scope will be an arrow, and we can write

> [the battery is flat $\rightarrow$ the starter will be dead, and you won't get the car started unless we push it] ***19.8***

Next we find the largest constituent sentence which has no truth-functor symbols in it:

> The starter will be dead, and you won't get the car started unless we push it. ***19.9***

This has the overall form 'ϕ and ψ', so that the truth-functor of largest scope in the translation of (19.9) will be conjunction:

> [the starter will be dead $\wedge$ you won't get the car started unless we push it] ***19.10***

The longest unanalysed constituent sentence of (19.10) is

> You won't get the car started unless we push it. ***19.11***

The overall form of (19.11) is 'ϕ unless ψ'. Remembering to remove the cross-referencing as we introduce '$\vee$', we write

> [you won't get the car started $\vee$ we'll push the car] ***19.12***

There remains

> You won't get the car started. ***19.13***
> $\neg$you'll get the car started.

The remaining sentences leave little opportunity to introduce more truth-functors; so at this point we call a halt and fit the pieces back together:

> [the battery is flat → [the starter will be dead ∧ [¬ you'll get the car started ∨ we'll push the car]]] ***19.14***

(19.14) is the correct analysis of (19.7).

After some practice, you will find you can write down an analysis such as (19.14) as soon as you see the original sentence. But for the moment you should go step by step, for safety.

You may find that you have to paraphrase quite freely in order to introduce truth-functors, just as in section 17. In some cases this may lead you to change the scope of a word. Two examples will make the point.

> He will be coming down by the 8.15 or the 9.15; if the former, then he will be in time to see the opening. ***19.15***
>
> [[he will be coming down by the 8.15 ∨ he will be coming down by the 9.15] ∧ [he will be coming down by the 8.15 → he will be in time to see the opening]]

Here we repeat part of the first half, to eliminate the cross-referencing ('the former'); this brings the part about 8.15 into the scope of 'if'.

> If you cut the party, then Jane, who is none too fond of you anyway, will just make life hell for you. ***19.16***
>
> [Jane is none too fond of you ∧ [you will cut the party → Jane will just make life hell for you]]

Here we take the clause after 'who' outside the scope of 'if', in order to turn the 'who' into '∧'.

Exercise 19. Analyse each of the following sentences as faithfully as possible, using truth-functor symbols:

1. I shall not fail to write to you.
2. He was gassed, not shot.
3. He was neither gassed nor shot.
4. Nobody will get any chocolate if Tracey screams again.

5. If the State schools lack adequate space, then the private schools, providing as they do an excellent education, ease the burden on the State's facilities.
6. If the private schools are socially top-heavy, then they are perpetuating social injustice; but in that case they cannot reasonably demand charitable status.

Sentence Tableaux

As promised on page 65, we shall now show that logical analysis leads to simpler tableaux. You should re-read section 10 quickly, to recall the main features of tableaux.

20. Sentence Tableaux

In order to construct a tableau, one has to be able to do two things, both of which demand concentration:

(i) given a sentence S, to describe the situations in which S is true, using only sentences shorter than S;
(ii) given a set of short sentences, to determine whether the set is consistent.

We shall see that when truth-functors are used, (i) becomes completely automatic, so that a machine could do it.

Consider first the truth-table for '∧', which we gave in section 17:

ϕ	ψ	$[\phi\wedge\psi]$
T	T	T
T	F	F
F	T	F
F	F	F

20.1

We can think of 'ϕ' and 'ψ' as standing for two sentences. As the table shows, there is precisely one case in which '$[\phi\wedge\psi]$' is true, namely the case where ϕ is true and ψ is true. Hence if '$[\phi\wedge\psi]$' occurs in a tableau, we can tick it and add ϕ and ψ at the bottom of each unclosed branch containing it:

$$\checkmark[\phi \wedge \psi]$$

$$\phi$$
$$\psi$$

20.2

(20.2) will be called the *derivation rule* for '$[\phi \wedge \psi]$'.

Next we consider the sentence '$\neg[\phi \wedge \psi]$'. According to the truth-table for '$\neg$' in section 17, this sentence is true just if '$[\phi \wedge \psi]$' is false. According to (20.1), there are just three kinds of situation in which '$[\phi \wedge \psi]$' is false; we can group these situations into two overlapping groups, namely those where ϕ is false and those where ψ is false. But of course ϕ is false just if '$\neg\phi$' is true, and likewise with ψ. Drawing all this together, we see that '$\neg[\phi \wedge \psi]$' is true precisely if either '$\neg\phi$' is true or '$\neg\psi$' is true. The *derivation rule* for '$\neg[\phi \wedge \psi]$' in tableaux is therefore a branching rule, as follows:

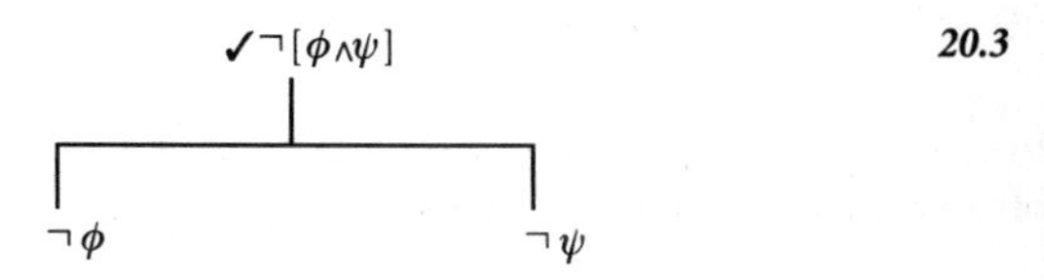

20.3

The derivation rules for the other types of complex sentence can be worked out from the truth-tables in just the same way. In fact this is so easy that you should try it at once, as an exercise.

Exercise 20A. Using the truth-tables given in section 17, and taking (20.2) and (20.3) as guides, work out what the derivation rules must be for the following forms of sentence:

1. $\neg\neg\phi$
2. $[\phi \vee \psi]$
3. $\neg[\phi \vee \psi]$
4. $[\phi \rightarrow \psi]$
5. $\neg[\phi \rightarrow \psi]$
6. $[\phi \leftrightarrow \psi]$
7. $\neg[\phi \leftrightarrow \psi]$

(You will find all the derivation rules, including the answers to this exercise, listed on pp. 191–2 and 281–2.)

To test the consistency of a set of sentences, we can now proceed as follows. First, analyse the sentences by means of symbolic truth-functors, just as in section 19. Then apply the tableau technique, using the derivation rules we have just described. A tableau which uses just these derivation rules is called a *sentence tableau.*

For instance, let's see how the example on p. 31 fares under this treatment. The set of sentences was

If cobalt but no nickel is present, a brown colour appears. **20.4**
Nickel and manganese are absent.
Cobalt is present but only a green colour appears.

After analysis, (20.4) becomes

[[cobalt is present ∧ ¬ nickel is present] → a brown colour appears] **20.5**
[¬ nickel is present ∧ ¬ manganese is present]
[cobalt is present ∧ only a green colour appears]

Using the derivation rules, we construct the tableau

✓ [[cobalt is present ∧ ¬ nickel is present] → a brown colour appears] **20.6**
✓ [¬ nickel is present ∧ ¬ manganese is present]
✓ [cobalt is present ∧ only a green colour appears]

¬ nickel is present.
¬ manganese is present.

Cobalt is present.
Only a green colour appears.

✓ ¬[cobalt is present ∧ ¬nickel is present] | A brown colour appears.

¬ cobalt is present. | ¬¬ nickel is present.

Nickel is present.

You should compare (20.6) carefully with (10.11).

Notice that if a sentence ϕ and its negation '$\neg\phi$' both occur in one branch, then the sentences in the branch are inconsistent, and we can close the branch. In fact we could have closed the middle branch of (20.6) as soon as we had written '¬¬ nickel is present', since this contradicts '¬ nickel is present' higher up in the branch.

Exercise 20B. Turn back to Exercise 10 (p. 36). Analyse each set of sentences there by symbolic truth-functors, and then test for consistency using sentence tableaux.

In section 11 we saw that the validity of an argument can be tested by checking the consistency of its counterexample set. The argument

$P_1 \ldots P_n$. *Therefore C.* **20.7**

is valid if and only if its counterexample set

$P_1 \ldots P_n$. 'It's not true that *C*.' **20.8**

is inconsistent. This means that sentence tableaux can be used to test the validity of arguments.

Here is an example. We shall test the validity of the argument

The mother will die unless the doctor kills the child. **20.9**
If the doctor kills the child, the doctor will be taking life.
If the mother dies, the doctor will be taking life.
Therefore either way, the doctor will be taking life.

We first analyse:

[the mother will die ∨ the doctor will kill the child] **20.10**
[the doctor will kill the child → the doctor will be taking life]
[the mother will die → the doctor will be taking life]
Therefore the doctor will be taking life.

The counterexample set is:

[the mother will die ∨ the doctor will kill the child] **20.11**
[the doctor will kill the child → the doctor will be taking
life]

[the mother will die → the doctor will be taking life]
¬ the doctor will be taking life.

We test the consistency of (20.11) by a sentence tableau:

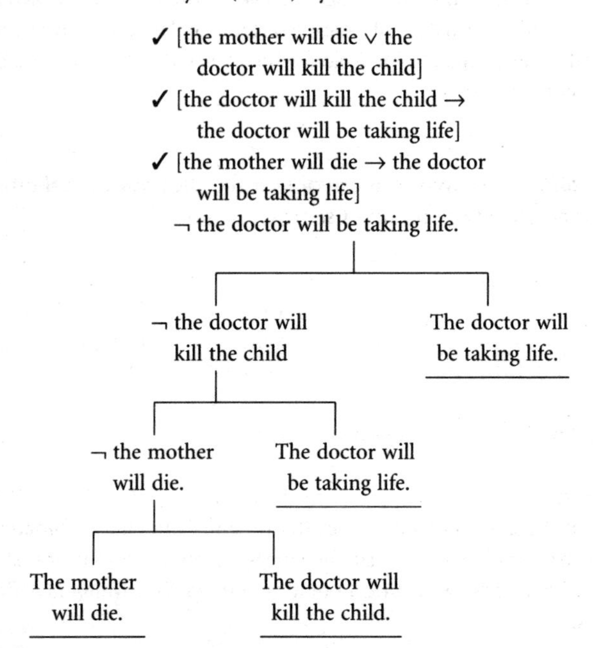

The tableau (20.12) is closed; therefore (20.11) was inconsistent, and so the argument (20.9) is *valid*.

Exercise 20C. Use truth-functors to analyse the sentences in the following arguments:

1. If the gunmen are tired, then they are on edge. If the gunmen are armed and on edge, then the hostages are in danger. The gunmen are armed and tired. *Therefore* the hostages are in danger.
2. If the driver was in control, then he passed the signal if and only if it was green. The driver passed the signal, although he was in control. Only if the electronics were faulty was the signal green. *Therefore* the electronics were faulty.

3. If the boy has spots in his mouth, then he has measles. If the boy has a rash on his back, then he has heat-spots. The boy has a rash on his back. *Therefore* the boy hasn't got measles.
4. Either the vicar or the butler shot the earl. If the butler shot the earl, then the butler wasn't drunk at nine o'clock. Unless the vicar is a liar, the butler was drunk at nine o'clock. *Therefore* either the vicar is a liar, or he shot the earl.

Exercise 20D. Using your answers to Exercise 20C, test the validity of these four arguments by means of sentence tableaux.

---

21. Interpretations

Section 20 introduced one labour-saving device; in this section we meet another.

The most irksome feature of sentence tableaux is that we have to keep writing out the same sentences time after time in different combinations. It would be sensible to use single letters as abbreviations for short sentences. This leads to the notion of an *interpretation*, which is defined as follows.

By an *interpretation* we mean a list of capital letters, in which each letter has a declarative sentence assigned to it. The same declarative sentence may be assigned to two different letters, but each letter must have just one declarative sentence assigned to it. For example, here is an interpretation, written as we shall usually write them:

P: Cobalt is present. ***21.1***
Q: Nickel is present.
R: Manganese is present.
S: A brown colour appears.
T: Only a green colour appears.

Using an interpretation, we can translate a complex sentence completely into symbols. The resulting string of symbols is called a *formula*.

A good procedure for translating into formulae is to put in the truth-functor symbols first, as in section 19, and then to use the interpretation to remove the remaining pieces of English. For example, we can start by translating

If cobalt but no nickel is present, a brown colour appears. ***21.2***

[[cobalt is present $\wedge \neg$ nickel is present] $\rightarrow$ a brown colour appears]. ***21.3***

Then we can use the interpretation (21.1) to contract (21.3) into the formula

$$[[P \wedge \neg Q] \rightarrow S] \qquad \textbf{\textit{21.4}}$$

If a set of sentences has been completely symbolized, we can test it for consistency by means of sentence tableaux, as in section 20; but there is one difference. Some types of inconsistency become hidden from view when the sentences are all symbolized. For example the two sentences

Schubert died at the age of thirty-one. ***21.5***
Schubert died at the age of sixty-eight.

are inconsistent. But symbolized, they would look something like this:

P. Q. ***21.6***

There is nothing in (21.6) to indicate that it is inconsistent.

We shall count a set of formulae as being 'obviously inconsistent' if and only if the set contains both a formula and its negation. Accordingly, in a completely symbolic tableau, *we close a branch if and only if there is a formula ϕ such that both ϕ and its negation '$\neg \phi$' are in the branch.*

To locate the 'hidden' inconsistencies in a finished tableau, we must inspect one by one the branches which are not closed. In each such branch we must take all the formulae which are either a single letter or the negation of a single letter, and we must use the interpretation to translate them back into English sentences; the resulting set of sentences can be checked directly for inconsistencies. Of course it may happen that the finished tableau is closed. In this case the original set of sentences is certainly inconsistent – there is no need to hunt for hidden inconsistencies.

To illustrate all this, we return to the example on page 35:

If cobalt but no nickel is present, a brown colour appears. ***21.7***
Either nickel or manganese is absent.
Cobalt is present but only a green colour appears.

Analysing, we have

[[cobalt is present ∧ ¬ nickel is present] → a brown colour appears] **21.8**
[¬ nickel is present ∨ ¬ manganese is present]
[cobalt is present ∧ only a green colour appears]

Translating into symbols by means of the interpretation (21.1), we reach the formulae

$$[[P\wedge\neg Q] \rightarrow S], [\neg Q\vee\neg R], [P\wedge T] \qquad \textbf{21.9}$$

We test the consistency of (21.9) by means of a sentence tableau:

21.10

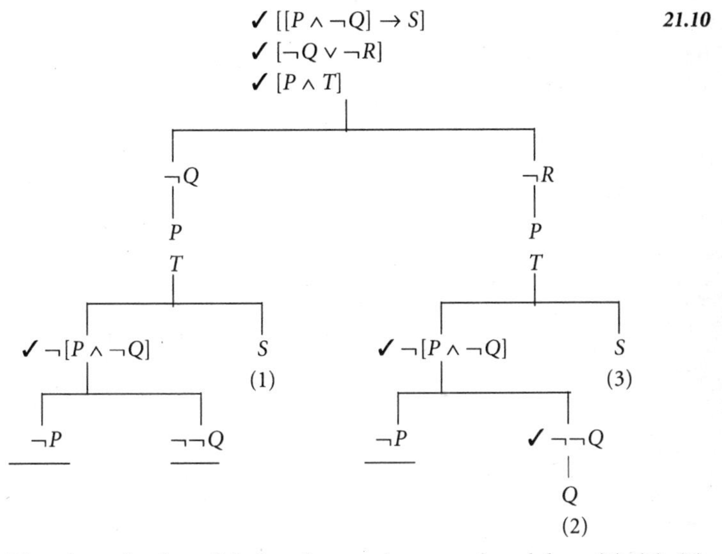

Three branches have failed to close; we have numbered them (1), (2), (3). If we translate the unticked formulae in each of these branches back into English by means of the interpretation (21.1), we find:

21.11

(1) Nickel is absent. Cobalt is present. Only a green colour appears. A brown colour appears.
(2) Manganese is absent. Cobalt is present. Only a green colour appears. Nickel is present.
(3) Manganese is absent. Cobalt is present. Only a green colour appears. A brown colour appears.

(1) is clearly inconsistent, although the branch is not closed; so we must discount it. The same applies to (3). There remains (2), which is consistent and so describes a situation in which (21.7) is true. Thus (21.7) has been proved *consistent.*

Exercise 21A. Using the following interpretation:

P: Income tax is levied on this income.
Q: The source of this income is employment.
R: The source of this income is a dividend liable to capital gains tax.
S: The source of this income is a dividend not liable to capital gains tax.
T: This income is attributable to government stocks.

translate each of the following sets of sentences into symbols:

1. The source of this income is not both employment and a dividend liable to capital gains tax. If income tax is levied on this income, then the source of this income is employment. The source of this income is a dividend liable to capital gains tax.
2. This income is attributable to government stocks if and only if its source is a dividend liable to capital gains tax. The source of this income is either employment or a dividend not liable to capital gains tax. Employment is not the source of this income, which is in fact attributable to government stocks.
3. Income tax is levied on this income if the source of this income is employment or a dividend not liable to capital gains tax. If the source of this income is a dividend liable to capital gains tax, then the income is attributable to government stocks. The source of this income is a dividend, but the income is not attributable to government stocks. Income tax is not levied on this income.

Exercise 21B. Which of the sets of sentences in Exercise 21A are consistent? Test by symbolic tableaux.

As in section 20, we can use these methods to test the validity of arguments. Here is an example.

> If the soil is suitable for carrots, then it is deep, sandy and free of stones. The soil is not suitable for linseed if it is *21.12*

sandy or a heavy clay. *Therefore* the soil is not suitable for both carrots and linseed.

A suitable interpretation would be:

P: The soil is suitable for carrots. ***21.13***
Q: The soil is suitable for linseed.
R: The soil is deep and free of stones.
S: The soil is a heavy clay.
T: The soil is sandy.

(Notice that in (21.13) we have saved ourselves a letter by taking 'deep' and 'free of stones' together; they never occur apart in (21.12).) By means of (21.13), we can translate (21.12) into

$$[P \to [R \wedge T]],\ [[T \vee S] \to \neg Q].\ \textit{Therefore}\ \neg[P \wedge Q]. \qquad \textbf{21.14}$$

Testing the validity of (21.14) by a tableau, we have

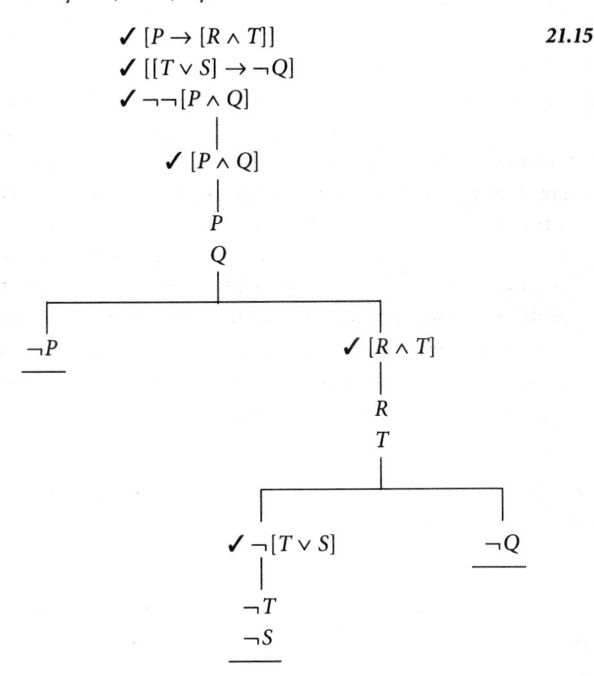

The tableau is closed. Therefore (21.12) was *valid*.

Exercise 21C. Test each of the following arguments for validity, using the interpretation provided. (For the origin of these examples, see the Answers.)

P: Uhha-muwas has bitten off Pissuwattis' nose.
Q: Pissuwattis is a female slave.
R: Pissuwattis is a free woman.
S: Uhha-muwas is liable to a 1 mina fine.
T: Uhha-muwas is liable to a 30 shekel fine.

1. If Uhha-muwas has bitten off Pissuwattis' nose, then he is liable to a 30 shekel fine. Uhha-muwas has not bitten off Pissuwattis' nose. *Therefore* Uhha-muwas is not liable to a 30 shekel fine.
2. If Uhha-muwas has bitten off Pissuwattis' nose, then, unless Pissuwattis is a female slave, Uhha-muwas is liable to a 1 mina fine. But Pissuwattis is not a female slave. *Therefore* either Uhha-muwas is liable to a 1 mina fine, or he has not bitten off Pissuwattis' nose.
3. Only if Pissuwattis is a free woman is Uhha-muwas, who has bitten off her nose, liable to a 1 mina fine. Pissuwattis is a female slave. *Therefore* Pissuwattis is not a free woman, and so Uhha-muwas is not liable to a 1 mina fine, even though he has bitten off her nose.
4. If Uhha-muwas has bitten off Pissuwattis' nose, then Uhha-muwas is liable to a 1 mina fine if Pissuwattis is a free woman, or a 30 shekel fine if Pissuwattis is a female slave. Uhha-muwas has bitten off Pissuwattis' nose, but is not liable to a 30 shekel fine. Pissuwattis is either a free woman or a female slave. *Therefore* Uhha-muwas is liable to a 1 mina fine.

Propositional Calculus

The task we now approach is that of *formalizing* logic. To formalize is to strip away the concepts which give meaning and application to the subject, so that nothing remains but bare symbols. Translation will disappear first, situations will fade away next, and finally even truth will make its exit.

On the negative side, this surgery pleases folk who regard possible situations as suspect anyhow. On the positive side, it exposes those parts of logic which can be developed as a mathematical theory. The mathematical theory we study in sections 22–25 is known as *propositional calculus*.

Unmathematical readers may prefer to leave out sections 22–25, and pass straight to section 26.

22. A Formal Language†

The first step in formalizing logic is to describe those strings of symbols which can occur as the translations of English declarative sentences. Our description must be purely formal, so that it must not mention English or translations. The simplest approach is to think of these strings as forming a language L_1, and to present a CF grammar for this language.

The CF grammar for L_1 shall be as follows:

$$\begin{array}{lll} 1 & \text{Fmla} \Rightarrow \neg\,\text{Fmla} \\ 2 & \text{Fmla} \Rightarrow [\ \text{Fmla} \wedge \text{Fmla}\ ] \\ 3 & \text{Fmla} \Rightarrow [\ \text{Fmla} \vee \text{Fmla}\ ] \\ 4 & \text{Fmla} \Rightarrow [\ \text{Fmla} \rightarrow \text{Fmla}\ ] \\ 5 & \text{Fmla} \Rightarrow [\ \text{Fmla} \leftrightarrow \text{Fmla}\ ] \end{array} \qquad \textbf{\textit{22.1}}$$

† Readers who dislike mathematics can omit this section.

6 Fmla ⇒ P Indx
7 Indx ⇒ Indx Indx
8 Indx ⇒ $_0$

The terminal symbols of this grammar are the brackets '[', ']', the truth-functor symbols '¬', '∧', '∨', '→', '↔', the letter 'P' and the subscript '$_0$'; the initial symbol is 'Fmla'. A typical phrase-marker is

22.2

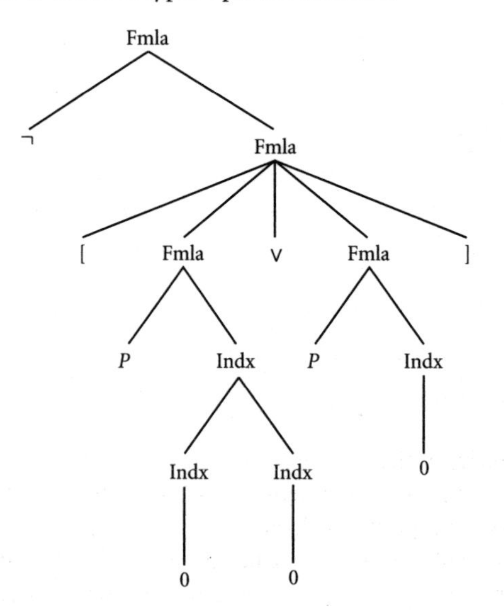

with terminal string '$\neg[P_{00} \vee P_0]$'.

The grammatical sentences of L_1 will be known as *formulae*. When a constituent of a formula ϕ is itself a formula, we say that the constituent is a *subformula* of ϕ. In particular every formula is a subformula of itself, since every formula is a constituent of itself (see p. 44). There are four subformulae of '$\neg[P_{00} \vee P_0]$', as shown in (22.3):

22.3

The simplest formulae are 'P_0', 'P_{00}', 'P_{000}', etc.; these are called *sentence letters*. Since long strings of subscripts are tiresome to write and read, we shall often allow ourselves to pretend that the sentence letters include '*P*', '*Q*', '*R*', etc.

Exercise 22. Which of the following are formulae of L_1? Write phrase-markers for those which are, and indicate their subformulae.

1. P_0
2. P_{00}
3. $[P_0 \rightarrow P_0]$
4. $[P_0 \vee P_{00}\ P_{000}]$
5. $[P_0 \wedge P_0 \wedge P_0]$
6. $\neg[\neg P_0]$
7. $P_0 \leftrightarrow P_{00}$
8. $[[\neg P_0 \rightarrow P_{00}] \rightarrow \neg\neg P_0]$

---

23. Truth-tables†

Having formalized the language, we now proceed to formalize the notions of situation and truth-value. This part of the subject is known as the *semantics* of propositional logic.

Suppose a sentence of English has been translated completely into symbols by means of truth-functors and an interpretation. The result is a formula of L_1; we may wish to determine the truth-value of this formula in some situation. The truth-values to be assigned to the sentence letters can be worked out by comparing the interpretation with the situation. Once the sentence letters have been tagged with truth-values, it's easy to calculate the truth-value of the whole sentence, by using the truth-tables of section 17. *The interpretation and the situation are needed only for assigning truth-values to sentence letters; truth-tables take care of the rest.*

An assignment of truth-values to sentence letters is called a *structure.* We shall use bold capital letters **A**, **B**, etc., to stand for structures. Here is an example of a structure:

P	*Q*	*R*
T	F	T

23.1

† Readers who dislike mathematics can omit this section.

(23.1) is the structure which assigns Truth to 'P' and 'R', and assigns Falsehood to 'Q'. We can list all the structures which make assignments to just the letters 'P', 'Q' and 'R'; there are eight of them:

P	Q	R
T	T	T
T	T	F
T	F	T
T	F	F
F	T	T
F	T	F
F	F	T
F	F	F

23.2

(23.2) shows the normal style for listing structures; 'T' comes above 'F', and the right-hand columns change faster than the left-hand ones. For one sentence letter there are two structures, for two sentence letters there are four, for four letters there are sixteen, and so on.

A formula ϕ of L_1 is said to be *defined in* a structure **A** if every constituent sentence letter of ϕ has a truth-value assigned to it by **A**. As we noted, a formula which is defined in **A** has a truth-value determined by **A** together with the truth-tables of section 17; we say that the formula is *true in* **A** if this truth-value is Truth, and *false in* **A** if the truth-value is Falsehood.

The truth-value of a formula in a structure can easily be calculated, by working step by step from shorter subformulae to longer ones. We write 'T' or 'F' as appropriate under each occurrence of a sentence letter, and for each complex subformula we record its truth-value under the truth-functor symbol of largest scope in the subformula. For example, we calculate the truth-value of the formula '$[P \rightarrow [Q \wedge \neg R]]$' in the structure (23.1) as follows. First we write

P	Q	R	$[P \rightarrow [Q \wedge \neg R]]$
T	F	T	T F T

23.3

The shortest complex subformula is '$\neg R$'; since 'R' is true, the truth-table for '$\neg$' on page 71 shows that '$\neg R$' must be false. We record this:

P	Q	R	$[P \rightarrow [Q \wedge \neg R]]$
T	F	T	T F FT

23.4

The next shortest subformula is '$[Q \wedge \neg R]$'; the truth-table for '$\wedge$' on page 72 shows that this must be false, since its two conjuncts are both false. We record this:

P	Q	R	$[P \rightarrow [Q \wedge \neg R]]$
T	F	T	T F F F T

23.5

The only remaining subformula is the whole formula itself. Using the truth-table for '$\rightarrow$' on page 74, we record its truth-value Falsehood beneath the truth-functor symbol '$\rightarrow$':

P	Q	R	$[P \rightarrow [Q \wedge \neg R]]$
T	F	T	T F F F F T

23.6

We can repeat this calculation for each of the structures listed in (23.2); each line of the following table represents one such calculation:

P	Q	R	$[P \rightarrow [Q \wedge \neg R]]$
T	T	T	T F T F F T
T	T	F	T T T T T F
T	F	T	T F F F F T
T	F	F	T F F F T F
F	T	T	F T T F F T
F	T	F	F T T T T F
F	F	T	F T F F F T
F	F	F	F T F F T F

23.7

The column written beneath '$\rightarrow$' shows the truth-value of the formula in each structure.

The table (23.7) is known as a *truth-table* for the formula '$[P \rightarrow [Q \wedge \neg R]]$'. It tells us the truth-value of this formula in every structure which makes assignments to just the sentence letters used in the formula. If a structure made assignments to some other letters as well, these would play no role in calculating the truth-value of our formula. For this reason, (23.7) in fact tells us the truth-value of the formula in every structure in which it is defined.

Exercise 23A. Give truth-tables for each of the following formulae. (Keep your answers; you will need them for Exercise 23B).

1. $[P \lor \neg P]$
2. $[[P \land Q] \lor [P \land \neg Q]]$
3. $[[P \to Q] \to [P \leftrightarrow Q]]$
4. $[[[P \to Q] \to P] \to P]$
5. $[[P \land Q] \land [\neg P \lor \neg Q]]$
6. $[[[P \to Q] \land [Q \to R]] \to [P \to R]]$

We can now formalize consistency. If X is a finite set of formulae of L_1, we write

$$X \models \qquad \textbf{23.8}$$

when there is no structure in which all the formulae of X are defined and true. (23.8) is read 'X is *semantically inconsistent*'; X is said to be *semantically consistent* when (23.8) is not true. The symbol '$\models$' can be pronounced 'semantic turnstile'.

Similarly we can formalize validity, by using the idea of counter-example sets as in section 11. If X is a finite set of formulae, and ψ a formula, we write

$$X \models \psi \qquad \textbf{23.9}$$

to mean that there is no structure in which ψ and all the formulae of X are defined, and all the formulae of X are true while ψ is false. (23.9) is read 'X *semantically entails* ψ'. Obviously

$$X \models \psi \text{ if and only if } X, \neg \psi \models. \qquad \textbf{23.10}$$

If X is empty, (23.9) boils down to

$$\models \psi \qquad \textbf{23.11}$$

which says that ψ is true in every structure in which it is defined; this formalizes the notion of a necessary truth. (23.11) is read 'ψ is a *semantic theorem*' or 'ψ is a *tautology*'.

Expressions of the forms (23.8), (23.9) and (23.11) are known as *semantic sequents*. Observe that semantic sequents are not themselves part of the language L_1, because '$\models$' is not a symbol of L_1. Rather, semantic sequents make statements about formulae of L_1. A semantic sequent is correct or incorrect, just as any statement is correct or incorrect. If a semantic sequent is incorrect, then there is a structure in which all the formulae of the sequent are defined, those to the left of '$\models$' are true, and any to the right of '$\models$' is false; such a structure is said to be a *counterexample* to the sequent. A sequent is correct if and only if it has no counterexample.

Truth-tables provide an easy check of the correctness of sequents. For example we shall test whether or not the following sequent is correct:

$$[P \to \neg Q], [P \leftrightarrow Q] \models \neg Q. \qquad \textbf{23.12}$$

To do this, we write out simultaneous truth-tables for these formulae. Recording only the columns which show the truth-values of the whole formulae, we have

P	Q	$[P \to \neg Q]$	$[P \leftrightarrow Q]$	$\models$	$\neg Q$
T	T	F	T		F
T	F	T	F		T
F	T	T	F		F
F	F	T	T		T

23.13

In the first structure listed in (23.13), '$[P \to \neg Q]$' is false. '$[P \leftrightarrow Q]$' is false in the second and third, while '$\neg Q$' is true in the fourth. Therefore (23.12) is *correct*.

We can check likewise whether

$$\models [P \to [Q \wedge \neg R]] \qquad \textbf{23.14}$$

The truth-table for this formula has already been calculated in (23.7). In fact this truth-table shows that the formula is false in the structure

P	Q	R
T	T	T

23.15

so it is not a tautology, and (23.14) is *incorrect*.

Exercise 23B. Which of the formulae of Exercise 23A are tautologies?

Exercise 23C. Use truth-tables to check which of the following semantic sequents are correct; indicate a counterexample to each incorrect one.

1. $P, [P \to Q] \models$.
2. $P \models [Q \to P]$.

3. $P, \neg P \models Q$.
4. $[Q \rightarrow P], [Q \rightarrow \neg P] \models \neg Q$.
5. $[P \rightarrow Q], [Q \rightarrow R] \models [R \rightarrow P]$.
6. $[R \leftrightarrow [P \vee Q]] \models [[R \wedge \neg P] \rightarrow Q]$.

We can summarize the connection between this section and earlier sections as follows.

Suppose a set of English sentences is translated, by means of a suitable interpretation, into a set X of formulae of L_1. If X$\models$, then the set of English sentences is inconsistent. For otherwise the sentences would be true together in some situation, and this situation together with the interpretation would give us a structure in which all the formulae of X are true.

Similarly, an argument which can be symbolized as a correct semantic sequent, with '$\models$' for '*Therefore*', must be a valid argument; likewise a sentence which can be symbolized as a tautology must be a necessary truth.

The converses fail. Obviously a set of sentences can be inconsistent for reasons which have nothing to do with truth-functors. Therefore the fact that such a set can't be symbolized as a semantically inconsistent set of formulae is no guarantee that the sentences are consistent. We shall return to this matter on page 121.

A List of Tautologies

1 $\neg [P \wedge \neg P]$
2 $[P \vee \neg P]$
3 $[P \leftrightarrow \neg \neg P]$
4 $[[P \vee Q] \leftrightarrow \neg [\neg P \wedge \neg Q]]$
5 $[[P \vee Q] \leftrightarrow [\neg P \rightarrow Q]]$
6 $[[P \vee Q] \leftrightarrow [Q \vee P]]$
7 $[[[P \vee Q] \vee R] \leftrightarrow [P \vee [Q \vee R]]]$
8 $[[P \vee P] \leftrightarrow P]$
9 $[P \rightarrow [P \vee Q]]$
10 $[P \rightarrow [Q \vee P]]$
11 $[[P \rightarrow R] \rightarrow [[Q \rightarrow R] \rightarrow [[P \vee Q] \rightarrow R]]]$
12 $[\neg P \rightarrow [[P \vee Q] \leftrightarrow Q]]$

13 $[[P \wedge Q] \leftrightarrow \neg [\neg P \vee \neg Q]]$
14 $[[P \wedge Q] \leftrightarrow \neg [P \rightarrow \neg Q]]$
15 $[[P \wedge Q] \leftrightarrow [Q \wedge P]]$
16 $[[[P \wedge Q] \wedge R] \leftrightarrow [P \wedge [Q \wedge R]]]$
17 $[[P \wedge P] \leftrightarrow P]$
18 $[[P \wedge Q] \rightarrow P]$
19 $[[P \wedge Q] \rightarrow Q]$
20 $[P \rightarrow [Q \rightarrow [P \wedge Q]]]$
21 $[[P \rightarrow Q] \rightarrow [[P \rightarrow R] \rightarrow [P \rightarrow [Q \wedge R]]]]$
22 $[P \rightarrow [[P \wedge Q] \leftrightarrow Q]]$
23 $[[P \wedge [Q \vee P]] \leftrightarrow P]$
24 $[[P \vee [Q \wedge P]] \leftrightarrow P]$
25 $[[P \wedge [Q \vee R]] \leftrightarrow [[P \wedge Q] \vee [P \wedge R]]]$
26 $[[P \vee [Q \wedge R]] \leftrightarrow [[P \vee Q] \wedge [P \vee R]]]$
27 $[[[P \vee Q] \wedge \neg P] \rightarrow Q]$
28 $[P \leftrightarrow [[P \wedge Q] \vee [P \wedge \neg Q]]]$
29 $[P \leftrightarrow [[P \vee Q] \wedge [P \vee \neg Q]]]$
30 $[[P \rightarrow Q] \leftrightarrow [\neg P \vee Q]]$
31 $[[P \rightarrow Q] \leftrightarrow \neg [P \wedge \neg Q]]$
32 $[P \rightarrow P]$
33 $[P \rightarrow [Q \rightarrow P]]$
34 $[[[P \rightarrow Q] \rightarrow P] \rightarrow P]$
35 $[[P \rightarrow [Q \rightarrow R]] \rightarrow [[P \rightarrow Q] \rightarrow [P \rightarrow R]]]$
36 $[[P \rightarrow Q] \rightarrow [[Q \rightarrow R] \rightarrow [P \rightarrow R]]]$
37 $[\neg P \rightarrow [P \rightarrow Q]]$
38 $[[\neg Q \rightarrow \neg P] \rightarrow [P \rightarrow Q]]$
39 $[[P \rightarrow [Q \rightarrow R]] \leftrightarrow [[P \wedge Q] \rightarrow R]]$
40 $[[P \leftrightarrow Q] \leftrightarrow [[P \wedge Q] \vee [\neg P \wedge \neg Q]]]$
41 $[[P \leftrightarrow Q] \leftrightarrow [\neg [P \wedge \neg Q] \wedge \neg [\neg P \wedge Q]]]$
42 $[[P \leftrightarrow Q] \leftrightarrow [[P \rightarrow Q] \wedge [Q \rightarrow P]]]$
43 $[P \leftrightarrow P]$
44 $[[P \leftrightarrow Q] \rightarrow [P \rightarrow Q]]$
45 $[[P \leftrightarrow Q] \leftrightarrow [Q \leftrightarrow P]]$
46 $[[P \leftrightarrow Q] \rightarrow [[Q \leftrightarrow R] \rightarrow [P \leftrightarrow R]]]$
47 $[[P \leftrightarrow Q] \leftrightarrow [\neg P \leftrightarrow \neg Q]]$
48 $[[P \leftrightarrow Q] \leftrightarrow \neg [P \leftrightarrow \neg Q]]$

24. Properties of Semantic Entailment†

When the shrubs are chopped away, the broader shapes of the landscape begin to appear.

Things which mathematicians prove are called *theorems*. Accordingly we shall list some theorems about semantic entailment and truth-tables, and see how they can be used, and what they tell us about validity of arguments in English.

I. *Extension Theorem*. If X and Y are finite sets of formulae, possibly empty, and ψ is a formula, then

$$\text{if } X \models \psi \text{ then } X, Y \models \psi. \qquad \textbf{24.1}$$

This is easy to see. If there is no structure in which all the formulae of X are true and ψ is false, then there can be no structure in which all the formulae of X and Y together are true and ψ is false. (The Extension Theorem tells us that a valid argument can't be made invalid by adding new premises – this is the so-called *monotonicity* property of validity.)

II. *Repetition Theorem*. If X is a finite set of formulae and ϕ is a formula in X, then $X \models \phi$.

This is obvious too; if every sentence of X is true in some structure **A**, and ϕ is in X, then ϕ must also be true in **A**. (As a statement about validity, this tells us that if the conclusion of an argument is one of the premises, then the argument is valid even if useless.)

III. *Cut Theorem*. If X is a finite set of formulae and ϕ and ψ are formulae, then

$$\text{if } X \models \phi \text{ and } X, \phi \models \psi, \text{ then } X \models \psi. \qquad \textbf{24.2}$$

The name refers to the fact that ϕ is 'cut out'. To see that the theorem is true, suppose that $X \models \phi$ and $X, \phi \models \psi$, and let **A** be any structure in which ψ is defined and all the formulae of X are true. We must show that ψ is true in **A**. ϕ may not be defined in **A**, but we can always add truth-value assignments to **A** so as to get a structure **B** in which ϕ is defined. Since $X \models \phi$, ϕ is true in **B**. Therefore, since $X, \phi \models \psi$, the formula ψ must also be true in **B**. But then ψ is true in **A** too, since **B** is only **A** with pieces added.

† Readers who dislike mathematics can omit this section.

IV. *Transitivity of* $\models$. If ϕ, ψ and χ are formulae, then

$$\text{if } \phi \models \psi \text{ and } \psi \models \chi, \text{ then } \phi \models \chi. \qquad \textbf{24.3}$$

Suppose $\phi \models \psi$ and $\psi \models \chi$. By the Extension Theorem, the latter implies that $\phi,\psi \models \chi$; this and the former imply that $\phi \models \chi$ by the Cut Theorem.

The next theorem needs some preparation. The idea is that the correctness of a semantic sequent never depends on the exact choice of letters in the sequent; for example we could replace '*P*' by '*Q*' throughout, without destroying correctness.

A *substitution scheme* is defined to be a list of sentence letters, to each of which a formula is assigned. For example, here is a substitution scheme, written as we shall write them:

$$\begin{array}{l} P \mapsto [R \to Q] \\ R \mapsto Q \end{array} \qquad \textbf{24.4}$$

(24.4) lists the sentence letters '*P*' and '*R*'; it assigns '$[R \to Q]$' to '*P*', and '*Q*' to '*R*'.

We apply a substitution scheme to a semantic sequent by taking each occurrence, in any formula of the sequent, of a sentence letter listed in the scheme, and replacing this occurrence by the formula assigned to this letter by the scheme. We do this simultaneously for all such occurrences throughout the semantic sequent. For example, if we apply (24.4) to the sequent

$$[P \to \neg Q], [P \leftrightarrow Q] \models \neg Q. \qquad \textbf{24.5}$$

the result is

$$[[R \to Q] \to \neg Q], [[R \to Q] \leftrightarrow Q] \models \neg Q. \qquad \textbf{24.6}$$

(The second part of (24.4) is not used, since *R* doesn't occur in (24.5).) When a substitution scheme is applied to a semantic sequent to produce a second semantic sequent, we say that the second sequent is a *substitution instance* of the first; thus (24.6) is a substitution instance of (24.5).

V. *Substitution Theorem*. Every substitution instance of a correct semantic sequent is again correct.

For suppose S_2 is the substitution instance of the sequent S_1 which comes from applying the substitution scheme T. If S_2 is not a correct sequent, then it has a counterexample **A**. Let **B** be the structure which is just like **A**, except that where a sentence letter is listed in T, **B** assigns to this letter the same truth-value as **A** assigned to the formula which T allots to the letter. Then **B** is a counterexample to S_1, so that S_1 is incorrect too.

For example, (24.5) happens to be a correct sequent, as we proved in (23.13). Therefore, by the Substitution Theorem, (24.6) is also correct. For another example, consider tautology 37 in the list on p. 105:

$$\models [\neg P \rightarrow [P \rightarrow Q]] \tag{24.7}$$

The Substitution Theorem tells us that '*P*' can be replaced in (24.7) by any formula, and '*Q*' by any formula. Thus the following are also tautologies:

$$[\neg [R \rightarrow Q] \rightarrow [[R \rightarrow Q] \rightarrow [P \vee S]]] \tag{24.8}$$
$$[\neg Q \rightarrow [Q \rightarrow P]]$$
$$[\neg Q \rightarrow [Q \rightarrow Q]] \text{ etc.}$$

Exercise 24A. Each of the following can be shown to be a tautology by applying a suitable substitution scheme to one of the tautologies in the list on pp. 104–5; say which tautology in the list, and give the substitution scheme.

1. $[[Q \rightarrow R] \vee \neg [Q \rightarrow R]]$
2. $[P \rightarrow [[P \rightarrow P] \rightarrow P]]$
3. $[[P \rightarrow [[P \rightarrow P] \rightarrow P]] \rightarrow [[P \rightarrow [P \rightarrow P]] \rightarrow [P \rightarrow P]]]$
4. $[[[Q \leftrightarrow R] \vee [P \wedge R]] \leftrightarrow [[[Q \leftrightarrow R] \vee P] \wedge [[Q \leftrightarrow R] \vee R]]]$

At first sight, the Substitution Theorem might seem to suggest that we can take any valid argument, and replace each occurrence of some sentence in the argument by another sentence (the same throughout), without destroying the validity. *This is true for those arguments which can be symbolized as correct semantic sequents of formulae of* L_1. For example, the valid argument

I am a mole and I live in a hole. *Therefore* I live in a hole. **24.9**

can be symbolized as

$$[P \wedge Q] \models Q \tag{24.10}$$

which is obviously a correct sequent. In this case we can replace 'I live in a hole' by any other declarative sentence, and the result will still be valid:

> I am a mole and diamonds consist of carbon. *Therefore* diamonds consist of carbon. ***24.11***

However, there are arguments which are valid for reasons not expressible in L_1. For these arguments, even a slight substitution may damage the validity beyond repair. For example, here is a valid argument:

> Suilven has always been unclimbable. *Therefore* nobody has ever climbed Suilven. ***24.12***

Slight changes yield the following argument:

> Indecent exposure has always been undesirable. *Therefore* nobody has ever desired indecent exposure. ***24.13***

(24.13) is hardly valid – indeed most people regard its premise as true and its conclusion as false.

It's an interesting psychological fact that we find (24.12) straightforwardly convincing and (24.13) completely unconvincing. It implies that the part of our mind involved in recognizing valid arguments doesn't work directly with the surface forms of the sentences involved. We must unconsciously carry out a kind of logical analysis on sentences before we think about the implications between them. Some people believe that this analysis is really the same process as finding the deep structures that we considered in section 12, though other people give reasons for doubting this. The question is hard, not least because it involves mental processes that are very fast and normally off the edge of consciousness. But experimental psychology moves forwards and it seems likely that we shall have a much better idea of the answer in fifty years' time. Meanwhile it's important to note that the notions of validity and consistency themselves, together with all the theorems about them in this section, are completely independent of any questions of psychology.

We say that the formula ϕ is *logically equivalent* to the formula ψ if $\phi \models \psi$ and $\psi \models \phi$. The following properties of logical equivalence are obvious, using the Repetition Theorem for VI and transitivity of '$\models$' for VIII:

VI. *Reflexiveness of Logical Equivalence.* Every formula is logically equivalent to itself.

VII. *Symmetry of Logical Equivalence.* If a formula ϕ is logically equivalent to a formula ψ, then ψ is logically equivalent to ϕ.

VIII. *Transitivity of Logical Equivalence.* If ϕ, ψ and χ are formulae, ϕ is logically equivalent to ψ and ψ is logically equivalent to χ, then ϕ is logically equivalent to χ.

By the definition, two formulae are logically equivalent if and only if they have the same truth-value in every structure in which they are defined. It follows that if two sentences can be symbolized as logically equivalent formulae, using the same interpretation, then the two sentences are true in exactly the same situations. The two sentences therefore play the same role as each other in any questions of consistency or validity. We should expect two logically equivalent formulae to behave the same way in all matters of logic. The next theorem illustrates this.

IX. *Congruence of Logical Equivalence.* If, in a correct semantic sequent, every formula is replaced by a formula logically equivalent to it, then the resulting sequent is also correct.

This is obvious from the truth-table test of the correctness of sequents. When the formulae are replaced by other formulae with the same truth-tables, the tables themselves are not altered.

For example, any formula logically equivalent to a tautology is itself a tautology.

If ϕ and ψ are formulae, and **A** is a structure in which both ϕ and ψ are defined, then '$[\phi \leftrightarrow \psi]$' is true in **A** precisely if ϕ and ψ have the same truth-value in **A**. This tells us that ϕ and ψ are logically equivalent if and only if

$$\models [\phi \leftrightarrow \psi]. \qquad \textbf{24.14}$$

Many of the tautologies listed on pp. 104–5 have the form '$[\phi \leftrightarrow \psi]$'; each such tautology provides us with a logical equivalence. For example, by tautology 16,

$[[P \wedge Q] \wedge R]$ is logically equivalent to $[P \wedge [Q \wedge R]]$. ***24.15***

By the Substitution Theorem, the letters 'P', 'Q' and 'R' in tautology 16 can be replaced by any other formulae, so that

for any formulae ϕ, ψ and χ, '$[[\phi \wedge \psi] \wedge \chi]$' is logically equivalent to '$[\phi \wedge [\psi \wedge \chi]]$'. ***24.16***

X. *Equivalence Theorem.* If ψ is any formula, ϕ a formula which occurs as a subformula of ψ, ϕ' a formula logically equivalent to ϕ, and ψ' a formula got from ψ by replacing one or more occurrences of ϕ in ψ by ϕ', then ψ' is logically equivalent to ψ.

This is most easily seen by imagining oneself working out the truth-tables of ψ and ψ'. From a certain point on, the calculations will be identical for the two formulae.

For example, according to tautologies 4, 31 and 41,

$$\begin{aligned} &\text{`}[\phi \vee \psi]\text{' is logically equivalent to `}\neg [\neg \phi \wedge \neg \psi]\text{',} \\ &\text{`}[\phi \rightarrow \psi]\text{' is logically equivalent to `}\neg [\phi \wedge \neg \psi]\text{',} \\ &\text{`}[\phi \leftrightarrow \psi]\text{' is logically equivalent to} \\ &\quad \text{`}[\neg [\phi \wedge \neg \psi] \wedge \neg [\neg \phi \wedge \psi]]\text{'.} \end{aligned} \tag{24.17}$$

Using (24.17) and the Equivalence Theorem, we can turn any formula of L_1 into a logically equivalent formula in which no truth-functor symbols occur except '$\neg$' and '$\wedge$'. To illustrate this, we take the formula

$$\neg [P \rightarrow [Q \vee R]]. \tag{24.18}$$

Since '$[Q \vee R]$' is logically equivalent to '$\neg [\neg Q \wedge \neg R]$', the Equivalence Theorem tells us that (24.18) is logically equivalent to

$$\neg [P \rightarrow \neg [\neg Q \wedge \neg R]]. \tag{24.19}$$

Since '$[P \rightarrow \neg [\neg Q \wedge \neg R]]$' is logically equivalent to '$\neg [P \wedge \neg \neg [\neg Q \wedge \neg R]]$', the Equivalence Theorem tells us that (24.19) is logically equivalent to

$$\neg \neg [P \wedge \neg \neg [\neg Q \wedge \neg R]]. \tag{24.20}$$

(24.20) has the desired form. We can improve it by using the fact (tautology 3) that '$\neg \neg \phi$' is logically equivalent to ϕ. Invoking this, we can bring (24.20) to the following logical equivalent of (24.18):

$$[P \wedge [\neg Q \wedge \neg R]]. \tag{24.21}$$

Exercise 24B. For each of the following formulae, find a logically equivalent formula in which '$\vee$', '$\rightarrow$' and '$\leftrightarrow$' do not occur.

1. $\neg [P \leftrightarrow Q]$

2. $[[P \vee Q] \vee R]$
3. $\neg [[P \vee Q] \rightarrow P]$

The fact that every formula of L_1 is logically equivalent to one in which '$\vee$', '$\rightarrow$' and '$\leftrightarrow$' don't occur has led some logicians to suggest that these three symbols should be dropped from the language, and only reintroduced as shorthands for the more complicated expressions given in (24.17). This would simplify the language L_1. Equally, we could use tautologies 5, 14 and a variant of 42 to drop '$\vee$', '$\wedge$' and '$\leftrightarrow$' in favour of '$\neg$' and '$\rightarrow$'.

One might ask how far we can go in the other direction. Can we find new truth-functors which can't be expressed in terms of the present ones? The answer is no.

XI. *Functional Completeness Theorem.* Any possible truth-functor which could be added to the language L_1 would only yield formulae logically equivalent to formulae which are already in the language.

This is proved as follows. Suppose we have an n-place truth-functor, which we shall write $f(\phi_1, \ldots, \phi_n)$. Since it is a truth-functor, it has a truth-table:

$\phi_1 \ldots \phi_n$	$f(\phi_1, \ldots, \phi_n)$
T . . . T	.
T . . . F	.
. .	.

24.22

where the column on the right lists the truth-values in all the structures listed on the left. We must find a formula of L_1 which has precisely the same truth-table as (24.22). We consider the possibilities. If there are no 'T's at all in the truth-table, then the formula '$[\phi_1 \wedge \neg \phi_1]$', which is always false, will serve. If there is precisely one 'T' in the column, then the following formula will serve:

$$[\ldots [\psi_1 \wedge \psi_2] \wedge \psi_3] \ldots \wedge \psi_n] \qquad \textbf{24.23}$$

where each ψ_i is ϕ_i or '$\neg\phi_i$' according as ϕ_i is true or false in the one structure which makes $f(\phi_1, \ldots, \phi_n)$ true. Finally if there is more than one 'T' in the column, then the formula

$$[\ldots [\chi_1 \vee \chi_2] \vee \chi_3] \ldots \vee \chi_k] \qquad \textbf{24.24}$$

will serve, where χ_1, χ_2, etc., are respectively the formulae (24.23) corresponding to the structures which make $f(\phi_1, \ldots, \phi_n)$ true.

For example, if a mad scientist invents the new truth-functor '$[\phi + \psi]$', with the truth-table

24.25

ϕ	ψ	$[\phi + \psi]$
T	T	F
T	F	T
F	T	T
F	F	F

then we can challenge his originality by producing a formula logically equivalent to '$[\phi + \psi]$', as follows. There are two 'T's in the column on the right of (24.25). The first is against the structure making ϕ true and ψ false; the corresponding formula (24.23) is

$$[\phi \wedge \neg \psi]. \qquad \textbf{24.26}$$

The second 'T' is against the third structure, which has the corresponding formula

$$[\neg \phi \wedge \psi]. \qquad \textbf{24.27}$$

Putting these together as in (24.24), we reach the desired formula

$$[[\phi \wedge \neg \psi] \vee [\neg \phi \wedge \psi]]. \qquad \textbf{24.28}$$

Exercise 24C. Construct a formula logically equivalent to the new formula $f(\phi, \psi, \chi)$ whose truth-table is

ϕ	ψ	χ	$f(\phi, \psi, \chi)$
T	T	T	T
T	T	F	F
T	F	T	F
T	F	F	T
F	T	T	T
F	T	F	F
F	F	T	F
F	F	F	F

XII. *Interpolation Theorem.* If ϕ and ψ are formulae such that $\phi \models \psi$, and at least one sentence letter occurs in both ϕ and ψ, then there is a formula χ such that

$$\phi \models \chi, \quad \chi \models \psi \tag{24.29}$$

and every sentence letter in χ is in both ϕ and ψ. (χ is known as an *interpolant* for the sequent '$\phi \models \psi$'.)

This is a little harder to prove. We say a structure **B** is an *expansion* of **A** if **B** is **A**, or consists of **A** together with assignments to other sentence letters not mentioned in **A**. To prove the Interpolation Theorem, one first lists all the structures which make assignments to just those sentence letters which are in both ϕ and ψ. Against each structure **A** one writes 'T' if **A** has an expansion in which ϕ is true, and 'F' if **A** has an expansion in which ψ is false. (This is possible since $\phi \models \psi$.) Any remaining structures are tagged 'T'. Using the procedure we employed to prove XI, one can construct a formula χ which has the truth-table thus defined. χ is the wanted interpolant.

For example,

$$[[P \rightarrow \neg Q] \wedge [Q \rightarrow P]] \models [\neg R \rightarrow [Q \leftrightarrow R]] \tag{24.30}$$

is a correct sequent. We write out the truth-tables:

24.31

P	Q	$[[P \rightarrow \neg Q] \wedge [Q \rightarrow P]]$
T	T	F
T	F	T
F	T	F
F	F	T

24.32

Q	R	$[\neg R \rightarrow [Q \leftrightarrow R]]$
T	T	T
T	F	F
F	T	T
F	F	T

There is one sentence letter used in both the formulae, namely 'Q'. According to the second line of (24.32), there is a structure in which 'Q' is true and '$[\neg R \rightarrow [Q \leftrightarrow R]]$' is false. According to the second or the fourth line of (24.31) there is a structure in which 'Q' is false and '$[[P \to \neg Q] \wedge [Q \to P]]$' is true. Therefore we set down

Q	*
T	F
F	T

24.33

Clearly the needed interpolant is '$\neg Q$'.

Exercise 24D. Find an interpolant for the sequent

$$[\neg [P \vee Q] \wedge [P \leftrightarrow R]] \models [[R \to Q] \wedge \neg [S \wedge R]]$$

The Interpolation Theorem says, roughly, that if a valid argument has a single premise which mentions sage but not onions, and a conclusion which mentions onions but not sage, then the argument can be split into two steps, and the half-way stage mentions neither onions nor sage. Perhaps this is obvious; but there is mathematical interest in the fact that if the argument can be symbolized as a correct semantic sequent of formulae of L_1, then so can the two half-way steps.

25. Formal Tableaux†

The next step in formalization is to remove the notion of truth altogether. We have seen that tableaux can be used to test consistency. Our procedure will be to redefine closed tableaux without any reference at all to truth, situations or structures. Once the definition has been given, we can compare closed tableaux with truth-tables and verify that a set of formulae generates a closed tableau if and only if the set is semantically inconsistent. But the definition itself will be completely independent of section 23.

Accordingly, we now define what we mean by an L_1-*tableau generated by* a finite set of formulae of L_1.

Let X be a finite set of formulae of L_1. Then by an L_1-*tableau generated by* X, we mean an array of formulae of L_1 arranged as an upside-down

† Readers who dislike mathematics can omit this section.

tree, with some branches possibly closed by having a line drawn below their bottom formula; so that (1) for each occurrence of a formula in the tree, either the formula itself is in X, or the occurrence is derived from another formula occurring higher in the branch according to one of the sentence derivation rules (as listed on pp. 281–2); (2) a branch is closed only if it contains an occurrence of some formula ϕ and an occurrence of its negation '$\neg \phi$'; (3) the top part of the tree consists of a list of all the formulae of X.

L_1-tableaux will be called simply *tableaux* for the rest of section 25. Note that we make no mention of ticks against the formulae; the ticks we used in earlier sections should be regarded as an aid to constructing a tableau, but not as a part of the tableau itself.

A tableau is said to be *closed* if all its branches are closed. If there is a closed tableau generated by the finite set X of formulae, then we say that X is *syntactically inconsistent*, and we express this by writing

$$X \vdash \qquad \textbf{\textit{25.1}}$$

('$\vdash$' is called the *syntactic entailment* symbol; it can be read 'syntactic turnstile'). By analogy with '$\models$', we also write

$$X \vdash \psi \qquad \textbf{\textit{25.2}}$$

to mean that X, $\neg \psi \vdash$ (where X is a finite set of formulae and ψ is a formula). (25.2) can be read as 'X *syntactically entails* ψ'. When X is empty, (25.2) contracts to

$$\vdash \psi \qquad \textbf{\textit{25.3}}$$

which can be read 'ψ is a *syntactic theorem*'.

Expressions of the forms (25.1), (25.2) and (25.3) are known as *syntactic sequents*. Like semantic sequents, they are used for making statements about formulae of L_1; they are not themselves formulae of L_1.

It can be shown – without appealing to extraneous notions such as truth or consistency – that if there is a closed tableau generated by X, then *every* tableau generated by X can be extended to a closed one. Because of this, we can check the correctness of a syntactic sequent simply by writing out a tableau generated by the appropriate formulae; the sequent is correct if and only if the tableau can be persuaded to close. For example we verify that

$$\vdash [P \rightarrow [Q \rightarrow P]] \qquad \textbf{\textit{25.4}}$$

as follows, by constructing a closed tableau generated by '$\neg [P \rightarrow [Q \rightarrow P]]$':

$$
\begin{array}{l}
\neg [P \rightarrow [Q \rightarrow P]] \\
\quad | \\
P \\
\neg [Q \rightarrow P] \\
\quad | \\
Q \\
\neg P \\
\hline
\end{array}
\qquad \textit{25.5}
$$

On the other hand we can show that

it is not correct that $\vdash [P \rightarrow [P \rightarrow Q]]$ ***25.6***

as follows:

$$
\begin{array}{l}
\neg [P \rightarrow [P \rightarrow Q]] \\
\quad | \\
P \\
\neg [P \rightarrow Q] \\
\quad | \\
P \\
\neg Q
\end{array}
\qquad \textit{25.7}
$$

There is no way of closing (25.7) or extending it with new formulae. This proves (25.6).

Exercise 25. Show that tautologies 9, 11, 16, 26 and 48 in the list on pp. 104–5 are syntactic theorems.

We shall now prove that a syntactic sequent is correct if and only if the corresponding semantic sequent is correct. Since

$X \models \psi$ if and only if $X, \neg \psi \models$ ***25.8***

and likewise

$X \vdash \psi$ if and only if $X, \neg \psi \vdash$ ***25.9***

we only need to prove that a finite set of formulae is syntactically inconsistent precisely if it is semantically inconsistent.

Accordingly we begin by proving that for every finite set X of formulae of L_1,

if X ⊨ then X ⊢. ***25.10***

(This is one of a cluster of theorems which go by the name of *Completeness Theorem*; the drift of all these theorems is that semantic inconsistencies can be detected by methods which were defined independently of the notion of truth.) We shall prove (25.10) by assuming that it is not true that X ⊢, and deducing that it is not true that X ⊨.

Assume then that X is a finite set of formulae of L_1, and it is not true that X ⊢. We can construct a tableau generated by X. Since no tableau generated by X is closed, there will certainly be a branch in our tableau which is not closed, and we can try to extend this branch downwards by applying the derivation rules and closing branches. Now as we keep closing branches and applying the derivation rules, we reach shorter and shorter formulae. This can't go on for ever; so we must eventually reach a point where any branch which can be closed is closed, and any formula which is added to the tableau by a derivation rule must already have occurred in the branch to which it is added. At this point we say the tableau is *finished*.

There must then be a finished but not closed tableau generated by X. We take such a tableau, and in it we choose a branch which is not closed. We construct a structure **A** as follows: **A** makes assignments to precisely those sentence letters which occur in formulae in the tree, and it assigns Truth to any sentence letter ϕ precisely if ϕ occurs as a formula in the chosen branch. We claim that every formula in the branch is true in **A**. This claim is proved by starting with the sentence letters and working up through longer and longer formulae. For example, if '¬ *P*' occurs in the branch, then '*P*' certainly doesn't occur in it, since otherwise the branch would have been closed; therefore '*P*' is false in **A**, and so '¬ *P*' is true in **A** by the truth-table for '¬'. As another example, suppose '[*P* → *Q*]' occurs in the branch; then since the tableau is finished, the derivation rule for such formulae shows that either '¬ *P*' or '*Q*' must occur in the branch. These formulae are shorter than '[*P* → *Q*]', so we can assume that if either of them is in the branch, it must be true in **A**. If '¬ *P*' is true in **A**, then '*P*' is false in **A**, and so '[*P* → *Q*]' is true in **A** by the truth-table for '→'; similarly if '*Q*' is true in **A** then '[*P* → *Q*]' is true in **A**. Either way, the formula '[*P* → *Q*]' is true in **A**. The arguments for the other types of formulae are similar.

We therefore have a tableau generated by X, a structure **A**, and a branch of the tableau which consists of formulae true in **A**. But the formulae of X are listed at the top of the tableau, so that they are in every branch. Hence all the formulae of X are true in **A**. This shows that it is not true that X $\models$. We have thus proved (25.10).

The whole proof above is simply a mathematical description of the way we used unclosed branches in section 10 to show that a set of sentences was consistent.

The converse of (25.10) must now be proved: if X is a finite set of formulae of L_1, then

$$\text{if } X \vdash \text{ then } X \models. \qquad \textit{\textbf{25.11}}$$

To prove this, we assume that X is semantically consistent, and we use this fact to enable us to find an unclosed branch in any tableau generated by X. The method is as follows. Since X is semantically consistent, there is a structure **A** in which all the formulae of X are defined and true. We shall try to walk downward through the tableau, starting at the top, so that every formula we pass through is true in **A**. When we reach the bottom edge of the tableau, we shall have paced out a branch consisting of formulae which are all true in **A**. If a formula ϕ is in the branch, then ϕ is true in **A**, so '$\neg\, \phi$' is not true in **A**, and hence is not in the branch. Therefore the branch is not closed.

It has to be checked that we can always keep going down the tableau, taking appropriate roads at the branching-points, until we reach the bottom edge of the tree. The first part of the journey is easy: at the top of the tree is a list of the formulae in X, and we know these are all true in **A**. From the bottom of the list onwards, we have to examine the next derivation rule which is used in the tree. For example, suppose the formulae traversed so far constitute a set Y, which is all true in **A**, Y includes the formula '$\neg\, [\phi \rightarrow \psi]$', and the next step in the tableau is to apply the derivation rule for this formula; then the next formulae in the tableau are ϕ and '$\neg\, \psi$'. But if '$\neg\, [\phi \rightarrow \psi]$' is in Y, then it's true in **A**, and so by the truth-table for '$\rightarrow$', ϕ and '$\neg \psi$' are both true in **A**. We can therefore extend our downward journey to take in these two formulae. For another example, suppose the set Y of formulae traversed so far contains the formula '$[\phi \vee \psi]$', and the derivation rule for this formula is the next one to be applied. Then since '$[\phi \vee \psi]$' is in Y, it must be true in **A**, and so by the truth-table for '$\vee$', at least one of ϕ and ψ is true in **A**. So we can continue our journey to take in the appropriate one of these two formulae.

This proves (25.11). (25.10) and (25.11) together show that semantic entailment and syntactic entailment amount to the same thing, although they were defined quite differently. With this our formalization is complete.

It must be admitted that there is an easier way to formalize this part of logic. One can simply regard truth-tables themselves as meaningless arrays of symbols constructed according to certain formal rules. There are several reasons why logicians prefer to formalize in terms of tableaux or some similar proof-calculus. One reason is that tableaux can be adapted to other parts of logic where the analogues of truth-tables can no longer be formalized. (We shall return to this in section 41.)

Some logicians prefer to avoid truth-tables for another reason which has nothing to do with formalization. According to these logicians, the notion of truth leads to too many philosophical perplexities – we have seen a few in sections 5–8. Instead, they say, we should take as basic the notions of *proof* or *entailment*, and build our logical apparatus on these. This is broadly the position of the Dutch school of *Intuitionists*, who reject truth-tables as a general method of logic, and accept tableaux only in a modified form.

Designators and Identity

There are inconsistent sets of sentences whose inconsistency can't be proved by symbolizing them and then applying sentence tableaux; likewise there are valid arguments whose validity is not demonstrable by this method. One example is:

> The boy's father is liable for the damage caused. I am the boy's father. *Therefore* I am liable for the damage caused.

When we have analysed how a sentence can be built up from constituents such as nouns and verbs, we shall be able to see what rules govern the validity of arguments like the example above. In due course these rules will be added to the rules of sentence tableaux.

26. Designators and Predicates

The time has come to look closer inside sentences, and find constituents which are smaller than sentences.

In section 5 we saw that the truth-value of a sentence in a situation depends closely on the references of parts of the sentence. There are certain types of phrase which are specially suitable for referring to things; we shall pick these phrases out and call them *designators.*

Four kinds of phrase will count as designators. These are proper names, non-count nouns, singular personal pronouns and definite descriptions – we shall define them in a moment. It may seem arbitrary to select just these four kinds of phrase; in fact the idea is simply to pick out a class which will suit our purposes in later sections.

First, there are *proper names.* These are the names which have become attached to particular things or people by special convention.

Many logicians regard them as the paradigm of designators. Examples are:

Birmingham, Pope Paul VI, NATO, The Kama Sutra, Betelgeuse, Rio Tinto Zinc, Tuesday, Beowulf, the Gulf Stream. *26.1*

We shall count proper names as English phrases, even though they may not all be in the dictionary.

The second type of designator is *non-count nouns*. A frame test (see p. 45) will define these: they always yield grammatical sentences when put in frames such as

I want some *x*. *26.2*
x is splendid. *26.3*
They have too much *x*. *26.4*

but (without changing their senses) they never yield grammatical sentences when put in either of

I want a *x*. (or: I want an *x*.) *26.5*
The *x*'s are lovely. *26.6*

(A perturbation counts as ungrammatical for the present test.) Examples are:

butter, copper sulfate, bacon, poverty, music, politics, intelligence, violence, moonlight. *26.7*

The things which are referred to by non-count nouns are only 'things' in a tenuous sense, but we shall allow them as things for logical purposes.

The third type of designator is *singular personal pronouns*:

I	**me**	*26.8*
you		
he	**him**	
she	**her**	
it		

The fourth type of designator is *definite descriptions*. These are singular noun phrases beginning with any one of

the	**my**	**Henry's**	*26.9*
this	**your**	**Birmingham's**	

that	**his**	**Tuesday's**
	her	(etc.)
	its	
	our	
	their	

Examples are:

the largest marrow of the season *26.10*
this bottle of distilled water
that thing you were telling me about Mary
my aunt in Surbiton
your back problem
his performance of Schubert's last sonata
Birmingham's drainage system
Odysseus' homecoming

Phrases beginning with words from the second or third columns of (26.9) could all be regarded as short for phrases beginning with **the**:

*the aunt of me in Surbiton *26.11*
*the back problem of you
the homecoming of Odysseus

Phrases of these four types are commonly used to name things. They have something else in common: they are all noun phrases. This is not an accident. It will be highly convenient to have all designators in one phrase-class, and there are at least two reasons why the class of noun phrases is the right one to seek them in. The first reason is that all proper names are noun phrases. (This seems to be a pure accident of language. One can easily imagine a dialect in which all proper names are adjectives: the Hodges household, yon Birminghamly town.) The second reason is that English is very adept at making noun phrases out of other kinds of phrase. For instance you may say, in a smelly place,

It reeks horribly in here. *26.12*

The word 'reeks' refers to the smell, although **reeks** is not a designator. But you could have expressed exactly the same thought by turning 'reeks' into a definite description:

The reek in here is horrible. *26.13*

What kinds of noun phrase have we excluded from being designators? There are two main kinds. First, we have excluded plural noun phrases, such as **Sacco and Vanzetti** or **my knees**. The objection to these is that they are normally used to indicate more than one thing at once.

Second, we have excluded singular noun phrases beginning with any of the following:

some	**any**	**every**	**no**	*26.14*
something	**anything**	**everything**	**nothing**	
somebody	**anybody**	**everybody**	**nobody**	
someone	**anyone**	**everyone**	**no one**	
each	**a**	**much**	**either**	
all	**one**	**little**	**neither**	

Thus we have excluded

every dog *26.15*
anything you say
each time he knocks
all the world
a stitch
some mustard
much ado
one way to cure wind
no self-respecting citizen

Some of these phrases could be used in settings where one might reasonably say they referred to something. But usually they serve a quite different purpose, as we shall see in section 34 below. For example, consider

I have never, in spite of much searching, found **one way to cure wind**. *26.16*

Obviously the bold phrase is not referring to one way to cure wind!

Exercise 26. Find all the designators which occur as constituents in the following sentences; say which of the four types each designator belongs to.

1. Ordinarily I think he does this simply because the ox, which has suffered from parasites throughout the day, gets relief from its back

being rubbed, but I was told that he may occasionally as he does so speak a few words to God or to the ghosts.

2. To this rule, Dr Jekyll was no exception; and as he now sat on the opposite side of the fire – a large, well-made, smooth-faced man of fifty, with something of a slyish cast perhaps, but every mark of capacity and kindness – you could see by his looks that he cherished for Mr Utterson a sincere and warm affection.

Here is a sentence in which two designators occur as constituents; the designators are bracketed:

[Jeeves] poured [the sherry]. ***26.17***

We can analyse such a sentence into its constituent designators

Jeeves, the sherry ***26.18***

together with the matrix which contained them:

x poured *y*. ***26.19***

The letters '*x*' and '*y*' which occur in (26.19) are called *individual variables*. They mark the holes where the constituent designators should go; in the terminology of p. 66, the occurrences of these variables in (26.19) are *free*. The letters '*z*', 'x_1', 'x_2', etc., will also be used as individual variables.

The matrix (26.19) is an example of a *predicate*. More precisely, a *predicate* is defined to be a string of English words and individual variables, such that if the individual variables are replaced by appropriate designators, then the whole becomes a declarative sentence with these designators as constituents.

Here are some predicates. As you read them you should think of appropriate designators for them; remember that you only have to make a declarative sentence, not necessarily a true one.

x loves a bit of night-life. ***26.20***

Angela was somewhat lacking in *y*. ***26.21***

Thiamine pyrophosphate is required for the prevention of *x* in birds. ***26.22***

x loves *y*. ***26.23***

James Thurber wrote *x* while he was staying at *y*. ***26.24***

x_1 added to x_2 makes x_3. ***26.25***

x gave it to *z*, who promptly gave it back to *x*. ***26.26***

Notice that we don't expect every choice of designators to turn a predicate into a declarative sentence. Inappropriate designators can easily lead to selection violations, as in

?The sherry poured Jeeves. ***26.27***

?Angela was somewhat lacking in John. ***26.28***

In later sections we shall be obliged to give truth-values to sentences such as (26.27) and (26.28); we shall take them to be false.

Predicates are classified by the number of different individual variables which have free occurrences in them. (Occurrences which are not free will occur for the first time in section 35; we can ignore them for the moment.) Thus (26.20)–(26.22) are *1-place* predicates. (26.19), (26.23) and (26.24) are *2-place*, while (26.25) is *3-place.*

Note that (26.26) is *2-place*, since only two individual variables have free occurrences in it, though one occurs twice. When an individual variable has two or more free occurrences in a predicate, this is understood to mean that the individual variable must be replaced by *the same designator at each free occurrence*. For example the holes in (26.26) can be filled to produce

The plumber gave it to the clerk, who promptly gave it back to the plumber. ***26.29***

They can't be filled to produce

The plumber gave it to the clerk, who promptly gave it back to Father McCloskey. ***26.30***

Cross-referencing by pronouns (see p. 14) can often have the same effect as repetition of variables. For instance, if we replace the second occurrence of '*x*' in (26.26) by 'him', we have the predicate

x gave it to *y*, who promptly gave it back to him. ***26.31***

If, as in (26.29), we now replace '*x*' in (26.31) by 'the plumber' and '*y*' by 'the clerk', then we reach the sentence

The plumber gave it to the clerk, who promptly gave it back to him. ***26.32***

(26.32) is not quite a paraphrase of (26.29), because there are female plumbers; but the difference in meaning is slight.

It's often convenient to think of declarative sentences as being 0-place predicates. In line with this idea, we shall sometimes use sentence variables to stand for predicates as well as for declarative sentences.

For convenience we shall also allow ourselves to say simply that a designator *occurs* in a sentence, when we mean that it *occurs as a constituent*.

The word **predicate** is often used by grammarians and philosophers in ways which are at variance with the definition we have given. For example, some people use the word to mean **property** or **quality**.

27. Purely Referential Occurrences†

Most designators can be used on their own, and not as part of a sentence. Thus we find the label **white spirit** on a bottle, the tag **J. Smith** on a coat, the title **The accumulation of capital** on a book, or the sign announcing **Anne Hathaway's cottage**. When a designator can be used *on its own* in a situation, so as to refer to something, we call that thing the *primary reference* of the designator.

One might think that designators have a straightforward role to play in language: namely to refer to their primary references. One might think, for example, that

the Queen *27.1*

refers to the Queen, whether it occurs on its own or in a sentence. But not so. For instance, the historian G. M. Trevelyan wrote in 1922 that

The Queen had in 1840 married Prince Albert of Saxe-Coburg-Gotha. *27.2*

Queen Victoria, to whom Trevelyan's sentence referred, was dead in 1901. She is not now the primary reference of the designator (27.1), nor was she when Trevelyan wrote. The primary reference of (27.1), now and in England, is Queen Elizabeth II.

In fact there are many roles a designator may play in a sentence, besides referring to its primary reference. Sometimes, as in (27.2), it refers to

† The topics in this section and the next, though important, are rather subtle. You may find it best to read these two sections fairly fast, and come back to them again later.

something else; sometimes it doesn't refer to anything at all. We shall want to distinguish those cases where a designator does straightforwardly refer to its primary reference. This demands a test. The test we shall use is by no means perfect, because it sometimes gives rather obscure results; but it will point in the right direction.

The test is as follows. Suppose a designator D occurs in a sentence. We try to rewrite the sentence in the form

> D is a person (thing) who (which, such that) . . . he (she, it) . . . *27.3*

with the designator brought out to the front. If there is such a sentence (27.3) which expresses the same as the original sentence, granted the assumption that D has a primary reference, then we say that the occurrence of D in the original sentence is *purely referential.* The purpose of this test is to bring the designator to a position where it must refer to its primary reference.

For example, the sentence

> Mr Hashimoto is Japanese. *27.4*

can be paraphrased as

> Mr Hashimoto is a person who is Japanese. *27.5*

Therefore the occurrence of 'Mr Hashimoto' in (27.4) is purely referential. This fits our intuition that in (27.4) this designator serves only to refer to its primary reference, namely Mr Hashimoto.

Similarly, the sentence

> Mr Hashimoto is the Managing Director of Kiki Products Inc. *27.6*

can be paraphrased (not very beautifully) as

> The Managing Director of Kiki Products Inc. is a person such that Mr Hashimoto is him. *27.7*

This shows that the occurrence of 'the Managing Director of Kiki Products Inc.' in (27.6) is purely referential.

For a third example, consider the sentence

> It isn't true that Mr Hashimoto is divorced. *27.8*

This doesn't mean the same as

> Mr Hashimoto is a person such that it isn't true that he is divorced. ***27.9***

because if there is no such person as Mr Hashimoto, then (27.8) is true while (27.9) is false. But if we take it as granted that there is such a person as Mr Hashimoto, then (27.8) and (27.9) do express the same thing. Therefore the occurrence of 'Mr Hashimoto' in (27.8) is purely referential.

Here are some examples of occurrences which are not purely referential. Such occurrences fall broadly into four main groups.

(i) *Space – time contexts.*

Here the reference of a phrase in the past or future, or at some other place in the universe, is relevant. An example is

> It's unprecedented for the Managing Director of Kiki Products to be Japanese. ***27.10***

Any attempt to paraphrase this as

> The Managing Director of Kiki Products is a person such that it's unprecedented for him to be Japanese. ***27.11***

would be most unhappy; the present Managing Director maybe always was Japanese, long before he took this post on. To make (27.11) mean the same as (27.10), we would have to invent a fictitious person, who is now the present Managing Director, was the previous Managing Director, and will eventually become the next Managing Director. No such shifty figure is involved in (27.10). Thus the occurrence of 'the Managing Director of Kiki Products' in (27.10) is not purely referential.

Or again,

> For Kazaks, the nearest office of Kiki Products is 2000 kilometres away in Moscow. ***27.12***

This clearly doesn't mean the same as

> The nearest office of Kiki Products is a place such that for Kazaks it is 2000 kilometres away in Moscow. ***27.13***

From where I'm writing, the nearest office of Kiki Products is a place in Milton Keynes in Buckinghamshire. Even for Kazaks, Milton Keynes is nowhere near Moscow. So the occurrence of 'the nearest office of Kiki Products' in (27.12) is not purely referential.

(ii) *Modal contexts.*

These talk about possibilities or necessities. An example is

> Mr Hashimoto can't help being the Managing Director. ***27.14***

There is no fair paraphrase of (27.14) along the lines of

> The Managing Director is a person such that Mr Hashimoto can't help being him. ***27.15***
> (OR: The Managing Director is a person whom Mr Hashimoto can't help being.)

For instance, the Managing Director actually is Mr Hashimoto, and obviously Mr Hashimoto can't help being himself, so (27.15) is true; but Mr Hashimoto could and should have resigned years ago, so (27.14) is false. Thus the occurrence of 'the Managing Director' in (27.14) is not purely referential.

Another example, involving a subjunctive conditional, is:

> If the Managing Director had been chosen by ballot, he would have been some flashy whizz-kid. ***27.16***

The sentence

> The Managing Director is a person who, if he had been chosen by ballot, would have been some flashy whizz-kid. ***27.17***

is hopeless as a paraphrase of (27.16). Mr Hashimoto is far too sage and unassuming to be a flashy whizz-kid, even if he had been chosen by ballot.

(iii) *Intentional contexts.*

These talk about beliefs, hopes, knowledge, and other facets of people's minds. (The name is from the Latin *intentio animae*, 'directing of the mind'.) An example is:

> Smith never realized that Mr Hashimoto was the Managing Director. ***27.18***

The attempted paraphrase

> The Managing Director is a person such that Smith never realized that Mr Hashimoto was him. ***27.19***

fails, because it implies that Smith failed to realize that Mr Hashimoto was Mr Hashimoto. Thus the occurrence of 'the Managing Director' in (27.18) is not purely referential.

(iv) *Quotational contexts.*

These talk about the actual words used by some speaker or writer. An example is:

> Mr Hashimoto likes to be referred to as the Managing Director. ***27.20***

The sentence

> ?The Managing Director is a person such that Mr Hashimoto likes to be referred to as him. ***27.21***

is of doubtful grammaticality. If it means anything at all, it hardly means the same as (27.20). Hence again, the occurrence of 'the Managing Director' in (27.20) is not purely referential.

Sometimes an ambiguity prevents us from saying whether an occurrence of a designator in a sentence is purely referential. For example, a woman who has divorced her first husband and then remarried may say

> My husband used to play in the town band. ***27.22***

The primary reference of her phrase 'my husband' is her present husband. If she means him, then the occurrence of this phrase in (27.22) is purely referential. If she means her previous husband, the occurrence is not purely referential. Like all structural ambiguities, this ambiguity can be overcome by rewriting the sentence – for example, with the words **my present husband** or **my then husband**.

Exercise 27. In each of the following sentences, is the occurrence of the bold phrase a purely referential occurrence? If it is not, classify the context as space-time, modal, intentional or quotational.

1. We suspect Abe is **her latest lover.**
2. If Abe is **her latest lover**, she's in trouble.
3. They accused Abe of being **her latest lover.**
4. Abe may be **her only lover.**
5. Abe will soon be **her only lover.**
6. Abe is **her latest lover.**
7. **Cyanide** is composed of carbon and nitrogen.
8. Siam changed its name to **Thailand.**
9. George wants to be **the first person to swim the Atlantic.**
10. It's impossible to see **Oxford** from Bletchley.
11. I've got this nagging feeling we forgot to turn off **the oven** when we left the house.
12. **The King** is dead. Long live **the King**!

When an occurrence of a designator is not purely referential, what is its function? We can begin to answer this if we follow the four-part classification given above.

In space-time contexts, the designator points to what would be its primary reference if we were at other times or places. In modal contexts, the topic is the primary reference which the designator has in some hypothetical situations. To use the language of section 8, these two types of context serve as *situation-shifters.*

In quotational contexts, a designator usually refers not to its primary reference, but to itself – or sometimes to itself as written or spoken in some particular way. This is clearest in such examples as

THE SACRED RIVER is printed in capitals. *27.23*

In this book we must leave on one side the intentional contexts. Neither of the treatments described above will account for them satisfactorily, and there is no general agreement on a better way of handling them.

28. Two Policies on Reference

Referential failure makes problems for logic; so do designator occurrences which are not purely referential. To leave a problem unsolved is hurtful. But there is a limit to the number of questions we can tackle at once, and frankly these problems about reference are not the most pressing. We shall therefore pass a Self-Denying Ordinance, and restrict ourselves (at least until the end of section 41) to arguments and sets of sentences which do not raise these problems.

Specifically, we shall make two assumptions.

First Assumption. We shall assume that if a designator has a purely referential occurrence in a sentence (from an argument or a set of sentences under discussion), then *the designator is understood to have a primary reference.*

For example, when the set of sentences or the argument under review contains a sentence such as

> Cinnamic acid is made by Perkin's reaction. ***28.1***

we shall leave out of account any possible situations in which there is no such thing as cinnamic acid. The effect is just the same as if we had added the sentence

> There is such a thing as cinnamic acid. ***28.2***

to the set of sentences, or to the premises of the argument. (On the other hand we would allow situations in which there is no Perkin. **Perkin's reaction** is fairly regarded as an unanalysable proper name, which applies to the reaction in question, whether or not there ever was such a person as Perkin.)

The restriction imposed by this First Assumption is not very arduous. Poets and St Anselm apart, people don't often use designators in purely referential occurrences, unless they take it as understood that the designators do have something to refer to.

Here is an example of the kind of argument which our First Assumption forbids us to consider:

> Irene never had any daughters. *Therefore* it's not true that Sigmund is the father of Irene's eldest daughter. ***28.3***

Here the point at issue is whether or not the designator **Irene's eldest daughter** does have a primary reference. We make nonsense of the argument if we begin by assuming that it does.

The argument (28.3) is clearly valid. Logicians have various ways of taking it into account. One approach is to paraphrase the conclusion in such a way that the offending designator no longer occurs; we shall describe how in section 38.

Second Assumption. We shall assume that *every occurrence of a designator in a set of sentences or an argument is purely referential.* We can allow exceptions where a designator is obviously irrelevant to questions of consistency or validity.

For example, we shall leave out of account such arguments as

> I once dreamed I was the mean annual rainfall in Mombasa. *Therefore* dreams do not consist of assorted memories. ***28.4***

The occurrence of 'the mean annual rainfall in Mombasa' in the premise is not purely referential. (The mean annual rainfall in Mombasa is forty-seven inches; I never dreamed I was forty-seven inches.) The argument (28.4) seems valid, though this would need careful justification.

Probably no general approach will codify more than a fraction of the valid arguments in which designators have occurrences which are not purely referential. We shall reconsider this question in sections 42–4.

29. Identity

One particularly important predicate is the 2-place predicate

> x_1 is one and the same thing as x_2. ***29.1***

This predicate is called *identity*; in symbols it's written '$x_1 = x_2$', and the symbol '=' is read 'equals'. A sentence got by putting designators in place of 'x_1' and 'x_2' in (29.1) is called an *equation*.

Various English phrases can be paraphrased by means of identity. For example:

> Everest **is** the highest mountain in the world. ***29.2***
> Everest = the highest mountain in the world.

> Cassius Clay and Muhammad Ali **are the same person.** ***29.3***
> Cassius Clay = Muhammad Ali.

> This **is none other than** the lost city. ***29.4***
> This = the lost city.

> Two plus two **equals** four. ***29.5***
> Two plus two = four.

The word **identical** is normally used in English to express close similarity rather than identity. For example, identical twins are not one and the same twin, and two women who are wearing identical dresses are not wearing one and the same dress.

Everything equals itself. This undeniable truth is sometimes known as the *Law of Identity*; according to taste, it is either the supreme metaphysical truth or the utmost banality. Since it is true always and everywhere, we can't deny it with consistency; any set of sentences which does deny it must be inconsistent.

From the Law of Identity we deduce the *Identity Rule: if D is a designator, then any set of sentences containing the sentence*

'¬ D = D' **29.6**

is inconsistent, (Warning: (29.6) comes from the sentence 'D = D' by adding '¬' in front. It doesn't have a constituent '¬ D'. In fact '¬ D' is not a designator, just as there is no town called Not Buenos Aires.)

The Identity Rule is based on our First Assumption from section 28, as follows. The designator D has a purely referential occurrence in (29.6), so it is assumed to have a primary reference. The Law of Identity says that this primary reference is equal to itself, so that the sentence 'D = D' must be true, contradicting (29.6).

The First Assumption tacitly adds

'There is such a thing as D.' **29.7**

to the set of sentences. Without the assumption, the Identity Rule would actually be incorrect. For example, in a situation where there is no Loch Ness Monster, the sentence

It's not true that the Loch Ness Monster = the Loch Ness Monster. **29.8**

is true; so in the absence of our assumption, (29.8) is entirely consistent. Some logicians insist that the Loch Ness Monster is equal to itself even if it doesn't exist, so that our First Assumption is not needed here. We have rejected this way of reading '='. But what these logicians have in mind could perhaps be expressed better by saying that the *legend* of the Loch Ness Monster is equal to itself; this legend certainly exists.

There is a second law governing identity: if $b = c$, then anything which is true of b is also true of c. This obvious truth is known as *Leibniz's Law*, in honour of the seventeenth-century philosopher and mathematician G. W. Leibniz, who made a study of identity.

Leibniz's Law tells us that certain arguments are valid. The arguments in question can be described as follows. *Let* D *and* E *be designators, and* ϕ *a declarative sentence, and suppose* ψ *is got from* ϕ *by replacing one or more occurrences of* D *in* ϕ *by occurrences of* E. *Then both the arguments*

ϕ, 'D = E'. *Therefore* ψ. **29.9**

ϕ, 'E = D'. *Therefore* ψ. **29.10**

are valid. We call this *Leibniz's Rule.*

For example, let D be 'Mr Helly', let E be 'the treasurer', and let ϕ be

Mr Helly has lost the petty cash. **29.11**

Then ψ must be

The treasurer has lost the petty cash. **29.12**

and (29.9) is illustrated by the argument

Mr Helly has lost the petty cash. Mr Helly is the treasurer. *Therefore* the treasurer has lost the petty cash. **29.13**

This argument is obviously valid.

Leibniz's Rule rests on our Second Assumption from section 28. Suppose S is a situation in which the two premises of (29.9) are true; then since 'D = E' is true in S, these two designators must have the same primary reference in S. Since the occurrences of D and E in ϕ and ψ respectively are assumed to be purely referential, ϕ must say the same thing about the primary reference of D as ψ says about the primary reference of E. These are one and the same primary reference; so by Leibniz's Law, if ϕ is true in S, then so is ψ.

What would happen if we dropped the Second Assumption? The following argument will demonstrate; it is invalid, although it looks very much like (29.13):

Hamish can name Hamish's brother. Hamish's brother is the man in the distance. *Therefore* Hamish can name the man in the distance. **29.14**

(Say, Hamish lives with his brother Tavish, but at this moment Tavish is impossible to recognize in the gloomy mountain mist.) In the conclusion of (29.14), the occurrence of the designator 'the man in the distance' is not purely referential. In fact it is clear that

> Hamish can name the man in the distance. ***29.15***

can't be paraphrased as

> The man in the distance is a person whom Hamish can name. ***29.16***

since (29.15) is false and (29.16) is true in the situation we imagined. (29.15) could more fairly be paraphrased as

> Hamish can answer the question 'Who is the man in the distance?' ***29.17***

which is a quotational context – as the quotation marks show. The Second Assumption is violated in (29.14), and so the application of Leibniz's Rule is fallacious.

(29.14) might well be called the *fallacy of education* – the assumption that students who have learned something under classroom conditions will recognize the same thing when they meet it again in the hustle of the outside world.

Similar fallacious applications of Leibniz's Rule have sometimes been used to discredit Leibniz's Law. Often the fallacy is hard to winkle out. For example, Figure 2 is a picture of me as I was a few decades ago. The baby in the picture is me. Nevertheless the baby has no beard, whereas I have a beard. This seems to contradict Leibniz's Rule. Further reflection

Figure 2

shows that this paradox is just the result of careless statement. The person portrayed in the picture really *does now* have a beard, because it's me; however, he *did not* have a beard at the stage of his life which the picture portrays. Straighten the tenses and the problem disappears.

Exercise 29. Use Leibniz's Rule to find valid arguments with the following as their premises. In some cases there are two or more correct answers: give as many as you can.

1. Six is greater than five. Three plus three equals six.
2. I am thy father's spirit. I scent the morning air.
3. Parasurama is Rama. Rama is Krishna.
4. Rama is Parasurama. Rama is Krishna.
5. The daughters of Lear disown Lear. Lear is the king.
6. Errol owns the gun that fired this bullet. This bullet is the bullet that killed Cheryl. The gun that fired the bullet that killed Cheryl is the murder weapon.
7. June is the third month after March. March is the ninth month after June.
8. The Son suffers. The Father is the same God as the Son.

Relations

There are many valid arguments which revolve around such phrases as **longer than** or **the same length as**. To bring these arguments under control, we need to find an appropriate analogue of truth-tables. Such an analogue exists in *relations*, which serve to tell us when we get a true sentence by filling the blank spaces in a predicate. Relations are more complicated than truth-tables, mainly because there are only two truth-values for a sentence to take, while there are indefinitely many different things available to be named by designators.

30. Satisfaction

As we saw in section 16, a truth-table tells us what sentences we can put into the holes of a truth-functor, so as to make it into a true sentence. There was no need to list the constituent sentences themselves – it sufficed to say what their truth-values must be. In the same way we now ask what designators we can put for the individual variables of a predicate, so as to produce a true sentence. It would be pleasant if the answer could be given in terms of the primary references of the designators, and we shall see that by and large this is possible.

The key notion is that of *satisfaction*, which is a relationship between things and predicates. Since the full definition of satisfaction is indigestible, we shall postpone it and begin with some typical examples instead.

Consider the 1-place predicate

x is at least two years old. ***30.1***

Things which are at least two years old are said to *satisfy* this predicate. Naturally this depends on the situation. A thing starts by being no years old, and failing to satisfy (30.1). In later situations it becomes one year old, and still fails to satisfy (30.1). Only in those situations where the thing is two years old or more does it satisfy (30.1).

Similarly, in any situation, a thing is said to *satisfy* the 1-place predicate

Angela was somewhat lacking in *y*. ***30.2***

if, in the given situation, it is true that Angela was somewhat lacking in the thing. For example, the situation might be that you are making a report about Angela's performance in a business negotiation. Suppose that Angela showed enthusiasm but no tact during the venture. Then in this situation, tact satisfies (30.2) but enthusiasm fails to satisfy it.

You can now answer Exercise 30A by analogy with (30.1) and (30.2).

Exercise 30A. The following chart claims to record the number of houses built per year per thousand of population, in each country and each year shown:

	1954	1960	1966
U.S.A.	9	7	6
U.S.S.R.	7	12	10
Sweden	8	9	12

Which of the three countries satisfied which of the following predicates in 1960?

1. *x* builds at least ten houses per year per thousand of population.
2. Six years ago, *x* built fewer than nine houses per year per thousand of population.
3. In six years' time, *x* will be building more than ten houses per year per thousand of population.

There is a similar notion for 2-or-more-place predicates, but now it becomes important to check that the right objects are steered to the right holes in the predicate. For this a convention is needed, and we shall adopt the following one. If n is a number greater than 1, and ϕ is an n-place predicate, then we do not define satisfaction of ϕ unless the individual variables in ϕ are numbered 'x_1', 'x_2', . . ., 'x_n'. A list of n things $T_1, T_2, \ldots, T_n$ will be written thus:

$$\langle T_1, T_2, \ldots, T_n \rangle \qquad \textit{30.3}$$

and described as an *ordered* n-*tuple*; the things T_1, T_2, etc., are described as the *terms* of this ordered n-tuple. Ordered 2-tuples $\langle T_1, T_2 \rangle$ are known as *ordered pairs*.

Consider the 2-place predicate

x_1 is larger than x_2. ***30.4***

An ordered pair is said to *satisfy* (30.4) if its first term is larger than its second term. For example, the following ordered pairs satisfy it (in the present situation):

⟨Canada, Luxemburg⟩, ***30.5***
⟨Los Angeles, Oslo⟩,
⟨the standard metre, the standard yard⟩,
⟨50, $4\frac{1}{2}$⟩.

The following ordered pairs do not satisfy it (again in the present situation):

⟨Canada, Canada⟩, ***30.6***
⟨Luxemburg, Canada⟩,
⟨Oslo, Los Angeles⟩,
⟨the standard yard, the standard metre⟩,
⟨$4\frac{1}{2}$, 50⟩

Similarly, consider the 3-place predicate

x_1 is composed of x_2 and x_3. ***30.7***

An ordered 3-tuple is said to *satisfy* (30.7) if its first term is composed of its second term and its third term. Thus

⟨water, hydrogen, oxygen⟩ ***30.8***
⟨salt, sodium, chlorine⟩

both satisfy (30.7) (in the present situation), whereas

⟨bread, sweat, sorrow⟩ ***30.9***

does not.

Exercise 30B. There are nine different ordered pairs whose terms are taken from the set of countries

U.S.A., U.S.S.R., Sweden

Write down all nine.

Exercise 30C. According to the chart in Exercise 30A, which ordered pairs of countries satisfied which of the following predicates in 1960?

1. x_1 builds more houses per year per thousand of population than x_2 does.
2. Six years ago, x_1 built at least as many houses per year per thousand of population as x_2 did.
3. x_1 will build fewer houses per year per thousand of population in six years' time than x_2 did six years ago.

A collection of things or people which is under discussion – for any reason – will be called a *domain*; the people or things in the domain will be called *individuals* (as in 'individual variable'). A collection of ordered n-tuples of individuals is called an n-*place relation* in the domain. For example, in Exercise 30C there was a domain consisting of the three countries U.S.A., U.S.S.R. and Sweden, and the answers to the three parts of Exercise 30C were 2-place relations in this domain. 2-place relations are commonly called *binary relations.*

It often proves convenient to regard a single individual as an ordered 1-tuple, so that a collection of individuals counts as a 1-place relation. The answers to the three parts of Exercise 30A were 1-place relations. (1-place relations are also known as *classes.*)

There are two main ways to describe a relation. The first is by listing the n-tuples which are in it.

For example, suppose we have invited Anne, Brenda, Clothilde, Ashok, Brian and Carl to a party, and we want to arrange them in pairs for a game. Then the domain consists of these six people. To pair them off, we can write down a collection of ordered pairs, say:

$$\{\langle \text{Anne, Brian} \rangle, \langle \text{Brenda, Ashok} \rangle, \langle \text{Clothilde, Carl} \rangle\}. \qquad \textit{30.10}$$

(30.10) is a binary relation. (It's normal to write curly brackets '{', '}' at the ends when a relation is specified by listing the ordered n-tuples in it.) The order in which we write the three ordered pairs of (30.10) is not significant; in fact (30.10) is just the same binary relation as

{⟨Clothilde, Carl⟩, ⟨Brenda, Ashok⟩, ⟨Anne, Brian⟩}. ***30.11***

However, (30.12) below is a different relation, because the individuals are changed around within the ordered pairs:

{⟨Brian, Anne⟩, ⟨Ashok, Brenda⟩, ⟨Carl, Clothilde⟩}. ***30.12***

Exercise 30D. Let the domain consist of the numbers 1 and 2; there are then four ordered pairs of individuals, namely ⟨1, 1⟩, ⟨1, 2⟩, ⟨2, 1⟩, and ⟨2, 2⟩. There are sixteen different binary relations in this domain; write down all of them. (They include the *empty* relation { }, which has no ordered pairs in it.)

The second way to describe an n-place relation is by naming an n-place predicate ϕ and a situation S; the relation is to consist of all the ordered n-tuples of individuals which satisfy ϕ in S. The relation is said to be *expressed* by the predicate in the situation.

For example, let the situation be as things actually are, and let the domain be the collection of all towns in the world. The predicate

x_1 is further east than x_2. ***30.13***

expresses a binary relation in this situation; the binary relation contains such ordered pairs as

⟨New Delhi, Boston⟩, ⟨Moscow, Paris⟩, ⟨Kyoto, Nairobi⟩. ***30.14***

Of course it contains an extremely large number of other ordered pairs too. This illustrates one of the main advantages of describing a relation by a predicate which expresses it: many relations are far too large to be listed in full.

Before we close this section, we should attempt a proper definition of satisfaction. The following will serve. Suppose ϕ is an n-place predicate in which the individual variables 'x_1', . . ., 'x_n' have free occurrences. Then the ordered n-tuple $\langle T_1, \ldots, T_n \rangle$ is said to *satisfy* ϕ in a situation S if it is possible to turn ϕ into a declarative sentence which is true in S, by replacing each free occurrence of 'x_1' by a purely referential occurrence of a designator whose primary reference in S is T_1, and similarly with 'x_2' and T_2, etc., to 'x_n' and T_n. The condition that the occurrences of designators should be purely referential is needed to ensure that the truth-value of the whole sentence depends only on the individuals $T_1, \ldots, T_n$, and not on the designators which are chosen to name them.

31. Binary Relations

Binary relations are so many and so diverse that one might despair of finding any pattern in them. We shall not despair. In this section we shall describe some of the simpler patterns one might hope to find among binary relations. Then we shall see, in section 32, that these simple patterns do emerge in the relations expressed by a wide range of 2-place predicates.

There is a handy way to picture binary relations. Suppose we have a domain D and a binary relation R in D. Then we draw one dot for each individual in D, and we enclose all these dots in a curve to represent D itself. For example, if D is the class of integers from 1 to 6 inclusive, then we draw

31.1

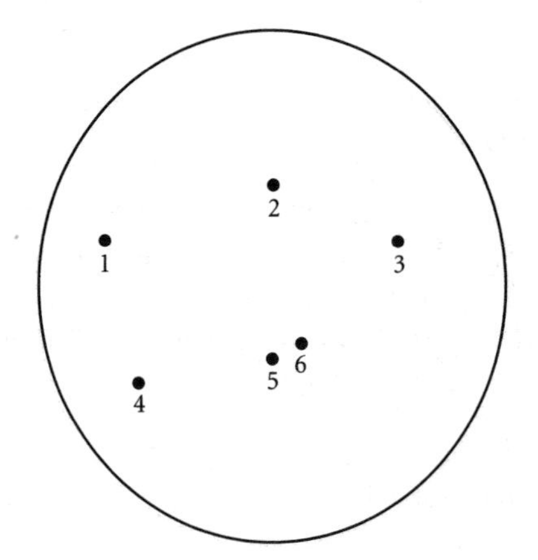

Now for each ordered pair $\langle b, c\rangle$ in R we draw an arrow

31.2

from b's dot to c's dot. If there are arrows going both ways:

31.3

we conflate them into one double arrow:

31.4

If R contains an ordered pair $\langle b, b\rangle$ whose first and second terms are the same individual, then there is an arrow from b to b, and conversely an arrow from b to b. So we draw a double arrow from b to itself (or rather from b's dot to itself):

31.5

This configuration (31.5) is called a *loop*. The whole diagram, loops and arrows and all, is called a *graph* of R.

To continue the example (31.1), suppose D is as before, and R is the relation

$$\{\langle 1, 2\rangle, \langle 3, 3\rangle, \langle 4, 4\rangle, \langle 4, 5\rangle, \langle 4, 6\rangle, \langle 5, 4\rangle, \langle 5, 5\rangle, \langle 5, 6\rangle\}. \quad \textbf{31.6}$$

Then this is a graph of R:

31.7

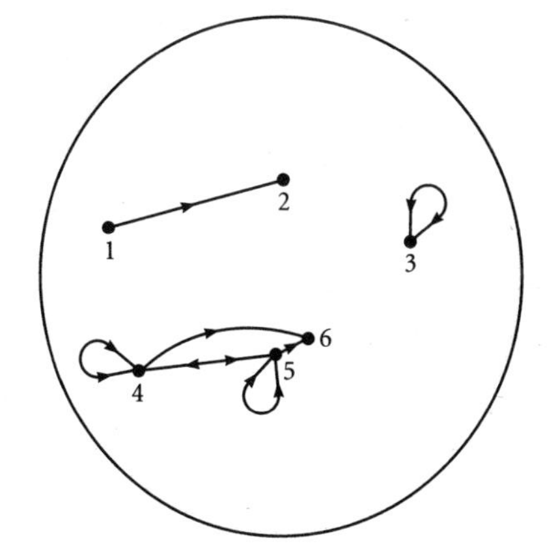

Of course this way of picturing binary relations is feasible only if D is finite, and fairly small at that. But if D is infinite, we can still imagine how an old Norse god with time on his hands might construct an infinite graph. So we have some excuse for classifying all binary relations in terms of their graphs.

(i) A binary relation is called

reflexive if every dot in its graph has a loop attached;
irreflexive if no dot in its graph has a loop attached;
non-reflexive if it's neither reflexive nor irreflexive.

For example, let the domain consist of the numbers 1, 2 and 3, and consider these three relations and their graphs:

$$A = \{\langle 1,1\rangle,\langle 2,2\rangle,\langle 2,3\rangle,\langle 3,3\rangle\};$$
$$B = \{\langle 1,2\rangle,\langle 2,3\rangle,\langle 3,1\rangle,\langle 3,2\rangle\};$$
$$C = \{\langle 1,1\rangle,\langle 3,2\rangle\}.$$

31.8

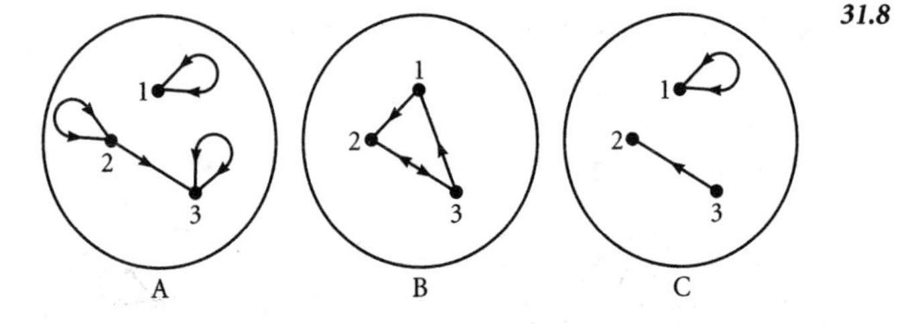

Here A is reflexive, B is irreflexive and C is non-reflexive.

We can apply this classification to predicates. For example, picking a word at random from the dictionary, we can consider the relation expressed by the predicate

x_1 dispurveys x_2. **31.9**

(in a given situation and domain). Regardless of what dispurveys means, this relation is reflexive if every individual **dispurveys** itself, and irreflexive if no individual dispurveys itself.

Exercise 31A. Let the domain be the class of all people now living, and let the situation be the present. Consider the binary relations expressed by the following predicates, and classify them as reflexive, irreflexive or non-reflexive:

1. x_1 is the same height as x_2.
2. x_1 is x_2's father.
3. x_1 laughs at x_2's jokes.
4. x_1 is taller than x_2.
5. x_1 is no taller than x_2.
6. x_1 is an environmentalist, and so is x_2.

(ii) A binary relation is called

symmetric if no arrow in its graph is single;
asymmetric if no arrow in its graph is double;
non-symmetric if it's neither symmetric nor asymmetric.

For example, let the domain consist of the numbers 1, 2 and 3, and consider these three relations and their graphs:

$$A = \{\langle 1, 1\rangle, \langle 2, 3\rangle, \langle 3, 2\rangle\};$$
$$B = \{\langle 2, 3\rangle\};$$
$$C = \{\langle 1, 1\rangle, \langle 2, 3\rangle\}.$$

31.10

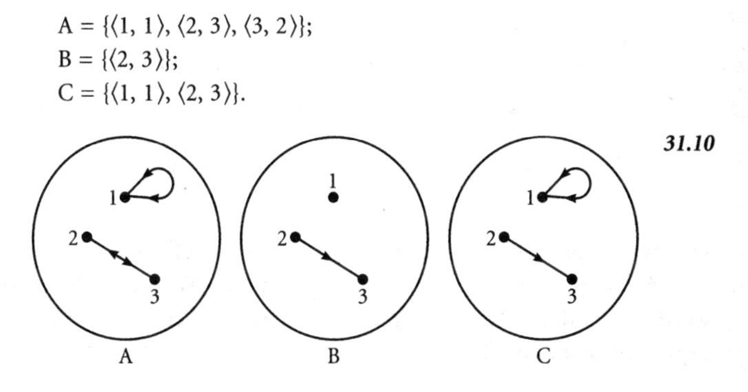

Then A is symmetric, B is asymmetric, and C is non-symmetric.

The relation expressed by (31.9) is symmetric if whenever an individual *b* dispurveys an individual *c*, then it is also true that *c* dispurveys *b*. The relation is asymmetric if whenever *b* dispurveys *c*, it is false that *c* dispurveys *b*.

Exercise 31B. Let the domain and situation be as in Exercise 31A. Consider the binary relations expressed by the following predicates, and classify them as symmetric, asymmetric or non-symmetric:

1. x_1 is younger than x_2.
2. x_1 is married to x_2.
3. x_1 is x_2's wife.
4. x_1 was x_2's best man.
5. x_1 and x_2 went to the same school.
6. x_1 is no younger than x_2.

Before we give our third classification, we must examine graphs once more. Suppose that there are dots representing *b*, *c* and *d* (not necessarily distinct individuals). We shall say there is a *broken journey* from *b* to *d* via *c* if there are an arrow from *b*'s dot to *c*'s dot, and an arrow from *c*'s dot to *d*'s dot. If there is a broken journey from *b* to *d*, we shall say this broken journey has a *short cut* if there is also an arrow direct from *b*'s dot to *d*'s dot. For example in the graph

31.11

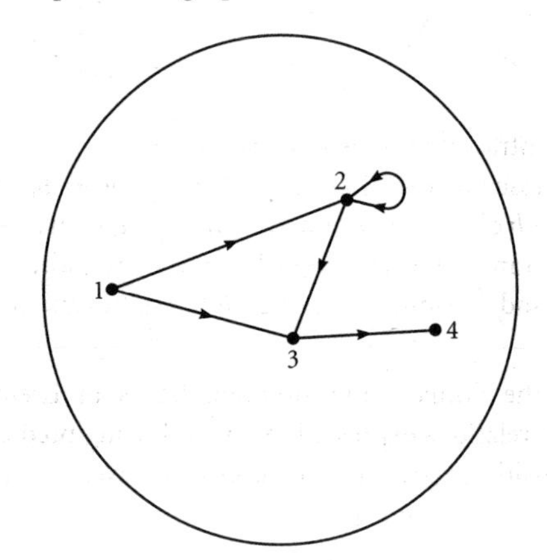

there are just six broken journeys:

from 1 to 2 via 2
from 1 to 3 via 2
from 2 to 2 via 2
from 2 to 3 via 2

} all with short cuts

from 1 to 4 via 3 }
from 2 to 4 via 3 } both without short cuts

(iii) A binary relation is called

{ *transitive* if its graph contains no broken journey without a short cut;
{ *intransitive* if its graph contains no broken journey with a short cut;
{ *non-transitive* if it's neither transitive nor intransitive.

For example, let the domain consist of the numbers 1, 2 and 3, and consider these three relations and their graphs:

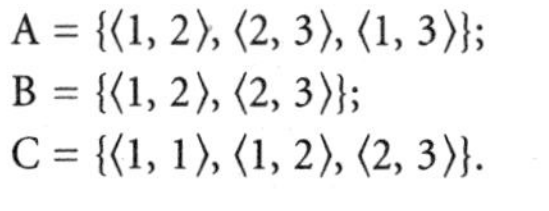

A = {⟨1, 2⟩, ⟨2, 3⟩, ⟨1, 3⟩};
B = {⟨1, 2⟩, ⟨2, 3⟩};
C = {⟨1, 1⟩, ⟨1, 2⟩, ⟨2, 3⟩}.

31.12

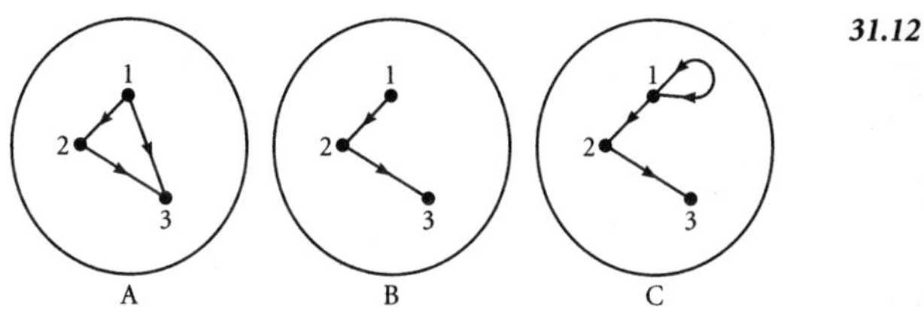

A is transitive, B is intransitive, C is non-transitive.

The relation expressed by (31.9) is transitive if whenever *b*, *c* and *d* are individuals such that *b* dispurveys *c* and *c* dispurveys *d*, then *b* dispurveys *d*. The relation is intransitive if whenever *b*, *c* and *d* are individuals such that *b* dispurveys *c* and *c* dispurveys *d*, then *b* doesn't dispurvey *d*.

Exercise 31C. Let the domain and situation be as in Exercise 31A. Consider the binary relations expressed by the following predicates, and classify them as transitive, intransitive or non-transitive:

1. x_1 knows x_2.
2. x_1 is descended from x_2.
3. x_1 is the father of x_2.
4. x_1 is no taller than x_2.
5. x_1 is married to x_2.
6. x_1 is a parent of x_2.

(iv) A binary relation is called *connected* if in its graph, any two different dots are connected by an arrow in one direction or the other (or both). For example, let the domain consist of the numbers 1, 2 and 3, and consider these two relations and their graphs:

A = {⟨1, 1⟩, ⟨1, 2⟩, ⟨2, 1⟩, ⟨2, 2⟩, ⟨2, 3⟩};
B = {⟨1, 2⟩, ⟨1, 3⟩, ⟨2, 3⟩, ⟨3, 1⟩, ⟨3, 3⟩}.

31.13

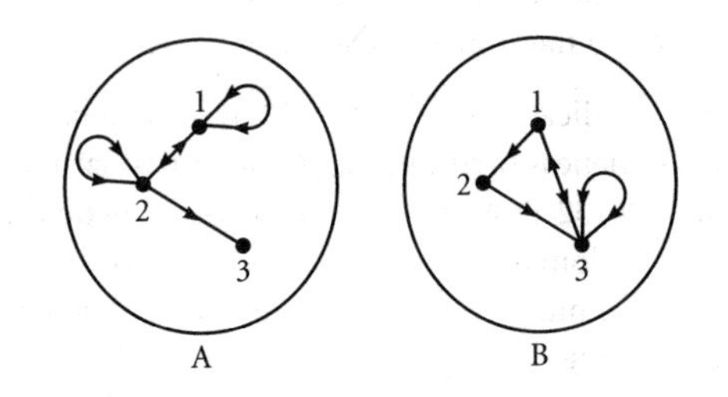

A is not connected, B is connected.

The relation expressed by (31.9) is connected if given any two different individuals, at least one of them dispurveys the other.

Exercise 31D. Let the domain and situation be as in Exercise 31A. Consider the binary relations expressed by the following predicates, and classify them as connected or not connected:

1. x_1 has the same surname as x_2.
2. x_1 is not the same person as x_2.
3. x_1 is no older than x_2.
4. x_1 is a farmer, while x_2 is not.

---

32. Same, at least and more

In this section we shall examine some kinds of predicate that have a logic of their own. Most adults find this logic all rather obvious once it is pointed out; but the psychologist Jean Piaget and his collaborators assembled impressive evidence that children find these matters hard to handle until age eleven or older.

(i) **Same**

Here are some typical 2-place predicates using the word **same**:

x_1 is the same height as x_2. ***32.1***
x_1 was born in the same town as x_2. ***32.2***
x_1 has exactly the same I.Q. as x_2. ***32.3***
x_1 has the same number of pages as x_2. ***32.4***
x_1 is one and the same thing as x_2. ($x_1 = x_2$.) ***32.5***

We shall refer to predicates like (32.1)–(32.5) as *sameness* predicates. In any situation, a sameness predicate carries with it a natural domain: the natural domain for (32.1) is the class of things or people which have a height, the natural domain for (32.4) is the class of books (i.e. things which have a certain number of pages), and so on. The natural domain for (32.5) is everything.

Every sameness predicate expresses a binary relation in its natural domain, in any situation. *The relation is always reflexive, symmetric and transitive*, apart from some few exceptions which we shall come to in a moment.

For example, if Chen was born in a town, then he was born in the same town as himself. If Chen was born in the same town as Ping, then Ping was born in the same town as Chen. If Chen was born in the same town as Ping, and Ping was born in the same town as Tsu, then Chen was born in the same town as Tsu. These sentences are true in any possible situation in which Chen, Ping and Tsu can be referred to – they are necessary truths (granted the existence of these people).

Exercise 32A. Propan-1-ol (P1 for short here) has the same empirical formula as propan-2-ol (P2 for short). Propan-2-ol has the same empirical formula as methoxyethane (M for short). What seven other facts of the same kind can you deduce immediately from these, without having to know any organic chemistry?

The exceptions we referred to are cases such as

x_1 went to the same school as x_2. ***32.6***
x_1 uses the same bank as x_2. ***32.7***

A person can have been to two different schools, and this allows (32.6) to express a relation which isn't transitive. If Jenny went with Albert to the Bellevue Infant School when she was five, and later went to the Crackworth High School with Roderick when she was fourteen, then we can't deduce that Albert went to the same school as Roderick, even though Albert went to the same school as Jenny and Jenny went to the same school as Roderick. In the same way a person can use two banks. This difficulty doesn't arise with most sameness predicates. For example, nobody has two different I.Q.s (unless something has gone wrong with the test).

(ii) **At least as, at most as**

x_1 is at least as red as x_2.	***32.8***
x_1 has at least as many children as x_2.	***32.9***
The half-life of x_1 is at most as long as the half-life of x_2.	***32.10***

We shall call predicates such as (32.8)–(32.10) *reflexive comparative predicates*. Like sameness predicates, they each have in any situation a natural domain; the domain for (32.8) consists of those things which it makes sense to describe as red; the domain for (32.10) consists of radioactive elements (i.e., things that have a half-life). The domain for (32.9) consists of those things or people that have some number of children – the number can be none.

Every reflexive comparative predicate expresses a binary relation in its natural domain, in any situation. *This relation is always reflexive, transitive and connected.*

For example, Pavel has at least as many children as Pavel has. If Pavel has at least as many children as Bengt, and Bengt has at least as many children as Giuseppe, then Pavel has at least as many children as Giuseppe. If Pavel and Giuseppe are two people, then either Pavel has at least as many children as Giuseppe, or Giuseppe has at least as many children as Pavel.

If we state a reflexive comparative predicate both ways round, the expressed relation is still reflexive and transitive; *but it becomes symmetric too.* For example, to symmetrize (32.8) we write

x_1 is at least as red as x_2, and x_2 is at least as red as x_1.	***32.11***

Likewise for (32.9):

x_1 has at least as many children as x_2, and x_2 has at least as many children as x_1.	***32.12***

These both-ways predicates can be shortened by using **exactly**:

x_1 is exactly as red as x_2. ***32.13***
x_1 has exactly as many children as x_2. ***32.14***

(32.13) and (32.14) can also be paraphrased as sameness predicates, by using words such as **number**, **amount** or **degree**:

x_1 has the same degree of redness as x_2. ***32.15***
x_1 has the same number of children as x_2. ***32.16***

Is this a general pattern? Can the symmetrized version of any reflexive comparative predicate be paraphrased as a sameness predicate? This would be surprising, because different nouns have to be used in different cases; the following would both be wrong:

*x_1 has the same number of redness as x_2. ***32.17***
*x_1 has the same degree of children as x_2. ***32.18***

We shall come back to this in section 33.

Exercise 32B. Try to find sameness predicates which paraphrase the symmetrized versions of the following reflexive comparative predicates:

1. x_1 is at least as small as x_2.
2. x_1 is at least as young as x_2.
3. x_1 is at most as far as x_2.
4. x_1 does at least as much as x_2.
5. x_1 breaks at least as often as x_2.
6. x_1's father is at least as affable as x_2's father.

(iii) **more, less, -er**

x_1 is cooler than x_2. ***32.19***
x_1 has more children than x_2. ***32.20***
x_1's bite is less dangerous than x_2's bite. ***32.21***

We shall call predicates such as (32.19)–(32.21) *irreflexive comparative* predicates. Each has a natural domain in any situation, just as reflexive comparative predicates do. In this domain it expresses a relation which is always *irreflexive, asymmetric and transitive.*

For example, the bite of the black widow spider (if it has one) is not less dangerous than the bite of the black widow. If the bite of the black widow is less dangerous than the bite of the tsetse fly, then the bite of the tsetse fly is not less dangerous than the bite of the black widow. If the bite of the black widow is less dangerous than the bite of the tsetse fly, and the bite of the tsetse fly is less dangerous than that of the lion, then the bite of the black widow is less dangerous than that of the lion.

Irreflexive comparative predicates are simply the back halves of reflexive comparative predicates, in the following sense. If we add **not** in a suitable way, we can always turn a predicate of the one sort into something which can be paraphrased as a predicate of the other sort. For example:

x_1 is not at least as red as x_2. ***32.22***
x_1 is less red than x_2.

x_1 is not at most as far as x_2. ***32.23***
x_1 is further than x_2.

x_1 is no cooler than x_2. ***32.24***
x_1 is at most as cool as x_2.

x_1's bite is not less dangerous than x_2's bite. ***32.25***
x_1's bite is at least as dangerous as x_2's bite.

There are several other connections between these types of predicate, as the following exercise illustrates.

Exercise 32C. Paraphrase each of the following predicates as a predicate of the type stated:

1. x_1 is at least as greedy as x_2 but x_2 is not at least as greedy as x_1. (*as an irreflexive comparative predicate*)
2. x_1 has at most as many antennae as x_2, but not the same number of antennae as x_2. (*as an irreflexive comparative predicate*)
3. x_1 is either knobblier than x_2, or precisely as knobbly as x_2. (*as a reflexive comparative predicate*)
4. x_1 is neither hotter than x_2, nor less hot than x_2. (*as a sameness predicate*)

---

33. Equivalence Relations

In this one section we shall intrude on the corner of logic known as the *theory of definition*, which studies the ways in which one concept can go proxy for others. I wish there were space for more; this is an unjustly neglected field.

A relation which is reflexive, symmetric and transitive (in a given domain) is called an *equivalence relation.*

An equivalence relation is easily recognized from its graph: the domain is split up into parts, and there is an arrow from b to c precisely when b and c are in the same part. (33.1) is an example:

33.1

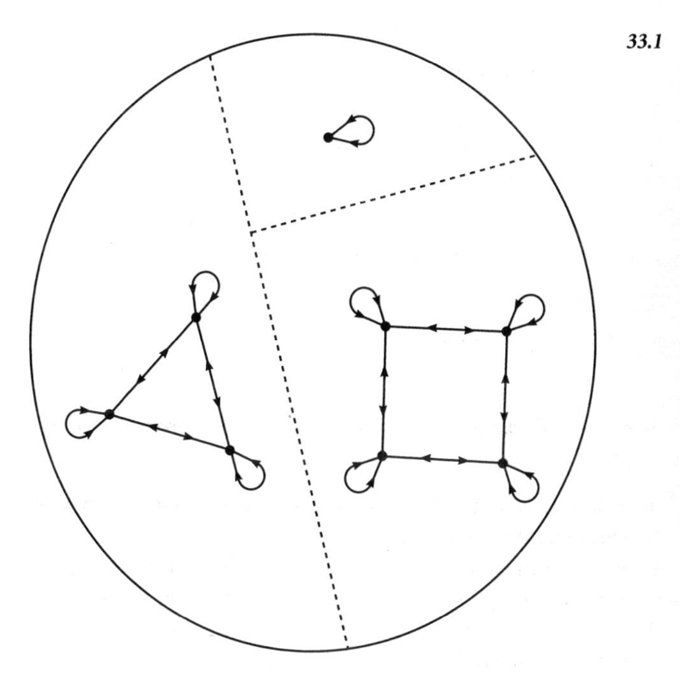

The parts are known as the *equivalence classes* of the relation; thus (33.1) has three equivalence classes.

In section 32 we saw that every sameness predicate expresses an equivalence relation (in a given situation). For example, the predicate

x_1 has the same number of children as x_2 **33.2**

expresses an equivalence relation. Two people are in the same equivalence class of this relation precisely if they have the same number of children. There is one equivalence class of people who have no children, another equivalence class of people who have just one child, and so on.

The converse is also true: not only does every sameness predicate express an equivalence relation, but *every equivalence relation is expressed by some sameness predicate.*

For suppose R is an equivalence relation; to express R, we simply write the predicate

x_1 has the same equivalence class of R as x_2. ***33.3***

(33.3) is a sameness predicate, and clearly it expresses R.

All this might seem a triviality, but in fact it is not. It tells us that we can introduce a certain kind of new noun into the language. If ϕ is any 2-place predicate which expresses an equivalence relation, then we can introduce into the language a new noun N, which can be taken to mean

equivalence class of the relation expressed by ϕ. ***33.4***

Then in place of ϕ, we can say

'x_1 has the same N as x_2.' ***33.5***

To describe this process, we call N an *abstraction* of ϕ.

For example, let ϕ be the predicate

x_1's mother was the same age when x_1 was born as x_2's mother was when x_2 was born. ***33.6***

This predicate expresses an equivalence relation in the domain of all human beings. Choosing an N to correspond, say **squirch**, we translate (33.6) into

x_1 has the same **squirch** as x_2. ***33.7***

Then **squirch** is an abstraction of the predicate (33.6). (Serious scientists would doubtless invent some more sober term, such as **maternal birth-age**.)

What do we gain by being able to abstract in this way?

Besides the easy point that N may be much shorter and simpler than ϕ was, the gains are two. First, it is obvious at a glance that (33.5) and (33.7) are sameness predicates, and hence that they express equivalence relations. This can serve as a mental guide in areas where arguments matter, for example in science. It may be much less obvious that ϕ itself expresses an equivalence relation.

For example, let ϕ be the predicate

x_1 is x_2, or could be exchanged for x_2 in the open market. ***33.8***

Under certain assumptions, which economists study, (33.8) expresses an equivalence relation. So we can abstract it to form the noun **exchange-value**. It's obvious that if b has the same exchange-value as c, and c has the same exchange-value as d, then b has the same exchange-value as d. The corresponding statement of transitivity using (33.8) is much less obvious.

The second gain is that other jobs may lie waiting for the new noun phrase N. For example, consider the predicate

x_1 holds at least as much as x_2. ***33.9***

This is a reflexive comparative predicate. Its symmetrized version

x_1 holds exactly as much as x_2. ***33.10***

expresses an equivalence relation. There's no need to invent an abstraction for (33.10); one already exists, namely **capacity**. But the word **capacity** is not only used to form the predicate

x_1 has the same capacity as x_2. ***33.11***

It also goes to work in such contexts as

This jug has a greater **capacity** than that one. ***33.12***

We want to increase the **capacity** of the stores. ***33.13***

The **capacity** of the tube is 5 c.c. ***33.14***

Note that each of these uses of **capacity** is closely related to the predicate (33.9) which first gave rise to the equivalence relation; the further uses of an abstraction noun rarely wander far from home. One doesn't debate whether a capacity is fluffy, or how old it is.

Incidentally this last example answers a question we posed on p. 153: can the symmetrized version of a reflexive comparative predicate always be paraphrased as a sameness predicate? The answer is that it can, because it expresses an equivalence relation, but a new noun may need to be invented for the purpose. We may well marvel that English has not yet evolved a uniform way of forming the required abstraction.

Science is riddled with abstractions of predicates which express equivalence relations. One example is **hardness** of minerals, as defined by the mineralogist F. Mohs: two minerals are said to have the same **hardness** if neither will scratch the other. Further examples are **species, blood-group, genotype, personality, temperature** and **valency**, to name a few.

There are dangers in abstraction. Gobbledygook is not the least. A more insidious danger is that the abstraction may be applied to things outside the original equivalence relation. For example, we noted that (33.8) expresses an equivalence relation under certain assumptions: these assumptions have to do with such things as the rationality of the merchants and the marketability of the goods. Even in a situation where the merchants are rational, highly perishable goods and goods of sentimental value may have to be excluded from the domain if (33.8) is to express an equivalence relation. If these goods are not in the domain, they do not have an exchange-value, and we land ourselves in contradiction if we pretend that they do have one. To insist on putting money values on everything is not only bad for the soul; it is also bad logic.

Quantifiers

According to the story, there was a man with a headache, who saw the advertisement

NOTHING ACTS FASTER THAN *****

– so at once he went and took nothing. This man failed to understand *quantifiers*. In the next few sections we hope to set him right.

34. Quantification

To *quantify* a predicate is to alter it so as to form a declarative sentence or a predicate in which fewer variables have free occurrences.

In English, the simplest way to quantify a predicate is to put a noun phrase in place of one of its individual variables. The noun phrase need not be a designator. In fact it can even be a plural phrase; but in this case we usually have to make some other adjustments to the predicate.

For example, the predicate

x is Japanese. ***34.1***

can be quantified to form

Somebody is Japanese. ***34.2***
Anyone who lives in Shimonoseki is Japanese. ***34.3***
Nothing marked 'Made in Britain' is Japanese. ***34.4***
One of the best contemporary poets is Japanese. ***34.5***
Most of the people who live in Kyoto are Japanese. ***34.6***
Three of these books are Japanese. ***34.7***
A few of the characters are Japanese. ***34.8***

Comparing (34.2)–(34.8), we can see that the added noun phrases all have a common purpose. Their purpose is to tell us something about *how many things of a certain type do or do not satisfy the predicate* (34.1). For example, in (34.5) we are told that at least one outstanding contemporary poet satisfies it; in (34.3) we are told that there are no people living in Shimonoseki who fail to satisfy it; and similarly with the rest.

Quantification is a complex matter. Piaget found that most children below the age of seven have difficulties handling even simple examples. To understand a sentence in which quantification occurs, we normally have to appreciate three things, namely the *profile* of the quantification, the *domain of quantification* and the *degree of exaggeration*. We shall discuss these in turn, starting with the profile.

Most English noun phrases can be written schematically in some such form as

the S, the Ss, an S, several Ss, a few Ss, half of the Ss, thousands of Ss, no S, (etc.) ***34.9***

For example **anyone who lives in Shimonoseki** can be written as **any S,** with 'person who lives in Shimonoseki' for S. Likewise **something** can be written as **some S,** with 'thing' for S.

We shall take as a schematic 1-place predicate:

x is a P. ***34.10***

If we put one of our schematic noun phrases (34.9) into the predicate (34.10), we get a sentence

the S is a P, the Ss are Ps, an S is a P, (etc.) ***34.11***

The sentence (34.11) then tells us something about the numbers of Ss which are or are not Ps (the hatched areas in (34.12)):

34.12

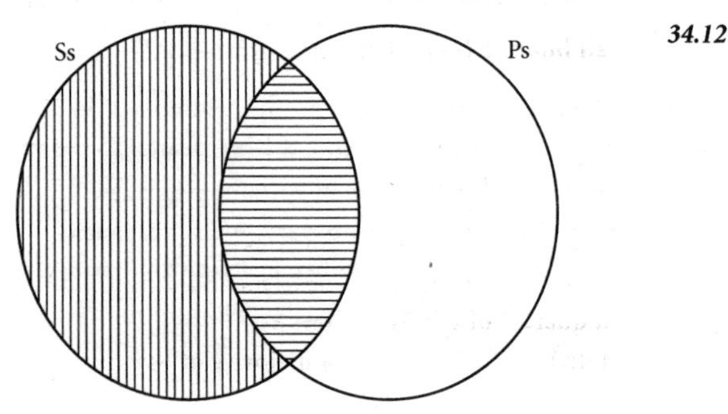

The type of information given about these numbers is called the *profile* of the quantification; it depends chiefly on the form of the noun phrase as shown in (34.9). The profiles which occur in English sentences usually fall into one of four groups.

In the first group of profiles, the sentence (34.11) tells us only about how many Ss *are* Ps (the horizontal hatching in (34.12)), but nothing about the Ss which are not Ps. The main examples are

no Ss				***34.13***
just one S	**at least one S**	**at most one S**	**one S**	
just two Ss	**at least two Ss**	**at most two Ss**	**two Ss**	
(etc.)	(etc.)	(etc.)	(etc.)	

some Ss, several Ss, many Ss, lots of Ss, a few Ss,
some of the Ss, several of the Ss, (etc.)

Two Ss is ambiguous; sometimes it's used to mean **just two Ss,** and sometimes **at least two Ss.**

In the second group of profiles, the sentence (34.11) tells us only about how many Ss *are not* Ps (the vertical hatching in (34.12)), and nothing about the Ss which are Ps. For example,

All Boy Scouts are honest. ***34.14***

tells us how many dishonest Boy Scouts there are – i.e. none. But it tells us nothing about how many honest Boy Scouts there are. (See section 7; we are assuming the weak reading.) Here are some other phrases which are used to indicate that no Ss are not Ps:

every S, **each** S, **all the** Ss, **every one of the** Ss, ***34.15***
each of the Ss, **the** Ss, **any** S.

Some other noun phrases with profiles of this second type are

all but one (**two, three,** etc.) **of the** Ss. ***34.16***

In the third group of profiles, the sentence (34.11) tells us *what proportion* of the Ss are Ps (relating the horizontal to the vertical hatching in (34.12)). Examples are:

half the Ss	**most Ss**	***34.17***
two thirds of the Ss	**nearly all Ss**	
a quarter of the Ss	**few Ss**	
(etc.)	**a majority of Ss**	

Note that **a few Ss** had the first type of profile, unlike **few Ss.**

The fourth group contains just one profile, namely that of definite descriptions. These are normally used only in situations where there is just one S. Thus it is wrong to say

Oleg is the tallest player in the team. ***34.18***

if two other players in the team are exactly as tall as Oleg, even if all the rest of the team are shorter.

Exercise 34. Both the following kinds of phrase occur sometimes with a profile from the first group, and sometimes with one from the second. Give examples to show this.

1. An S. 2. Ss.

We turn to the *domain of quantification*, and we begin with an example. Imagine this situation: the fog at London Airport is heavy and is not expected to lift for twenty-four hours. All planes have been grounded, and passengers with urgent business are advised to fly from Manchester instead. The staff at London Airport have put up notices which say

ALL FLIGHTS ARE CANCELLED ***34.19***

In this situation, (34.19) is true. But why is it true, if in fact planes are flying from Manchester? Don't flights from Manchester count?

The answer is that they don't, for purposes of a notice such as (34.19) in London Airport. 'All flights' in this notice refers only to flights from London Airport, and flights from any other airport simply do not count. The flights which count as relevant to the truth of (34.19) are said to be *in the domain of quantification*; those which don't count are not.

For another example, consider the cricket captain who is telling someone about his team; he says

Most of the players bowl right-handed. ***34.20***

The only players who count as relevant are the players in his team. These players are in the domain of quantification; the women in the badminton team down the road, who never bowl either right-handed or left-handed, are not.

In both these examples we can say that certain things definitely are in the domain of quantification, while certain other things definitely are not. For all the rest of the things in the world, it is quite arbitrary whether we choose to put them in the domain or not. In example (34.19) we can put the airport clocks into the domain or leave them out as we please, since they make no difference to the truth of (34.19) in the given situation. In example (34.20) we can toss a coin over the question whether the captain's dog is in the domain; it is not a player, so it has no effect on the truth-value of the given sentence. Because the domain of quantification is arbitrary to this extent, we have to admit that it is something of an artificial construct.

The domain of quantification depends on the situation; it also depends on the precise wording of the noun phrase. To see this, contrast the following two sentences:

Each golfer has a favourite golf course. *34.21*

All golfers have a favourite golf course. *34.22*

One would use (34.21) when a particular group of golfers was under discussion; only these golfers are in the domain of quantification. With (34.22) on the other hand, all golfers whatever must be included in the domain of quantification, even if the topic of conversation is one small group of golfers. The sentence

Every golfer has a favourite golf course. *34.23*

leaves it in doubt whether the domain is large or small – the context must decide.

The word **the** is often used to contract the domain to a group of people or things which have been already mentioned. This is the role of **the** in the sentence

All the golfers have a favourite golf course. *34.24*

which could be used to mean the same as (34.21). To contract the domain still further, one deploys phrases such as **all these golfers** or **all those golfers**.

Finally we must consider *exaggeration*; this is something which chiefly infects phrases such as (34.15), in the second group of profiles. In daily discourse, everybody uses these phrases in ways which aren't meant to be taken literally; since everybody does it, nobody is deceived.

For example, it is common practice to say 'every S', when one means 'every S, leaving aside some exceptions which make no difference to the point at issue':

Everybody has heard of Mao Tse-Tung. ***34.25***

(Infants? recluses?) Likewise one says 'Ss' when one means 'normal Ss', or even 'typical Ss':

Sparrows have a cheerful grey crown. ***34.26***

(Even filthy ones?)

Women are less interested in sport than their men-folk. ***34.27***

(All of them??)

Some writers on logical themes have taken a curiously strong moral stand against this kind of exaggeration. For example, Susan Stebbing, complaining about the exaggerated use of **everybody**, writes:

> There are serious dangers in indulging in such a habit . . . It encourages us to turn aside from contrary evidence, to oversimplify important issues, to attribute to other people an unwarranted extension of what they have been asserting.†

She goes on to argue that this type of verbal usage is a threat to political moderation. Maybe. When Susan Stebbing wrote, there were tyrannical regimes in Germany and Russia, and the Second World War was about to break loose; one can hardly blame her for invoking logic on the side of the angels. But a calmer assessment would say that we use language for many purposes, and different purposes call for different degrees of precision. Too little precision and you deceive people; too much precision makes you a pedant.

35. **All** and **some**

Logicians cultivate three main methods of quantification. The first is to put a designator in place of a variable; this is called *instantiation*. The remaining two are called *universal quantification* and *existential quantification*. They are both a little different from anything which occurs in

† *Thinking To Some Purpose*, Penguin, 1939, p. 125.

ordinary English; but we shall see that they can be used to express the sense of many different English noun phrases.

(i) *Universal quantification*

Suppose we have a predicate, written schematically as

$$- - - - x **** , \qquad \textit{35.1}$$

in which the individual variable '*x*' has one or more free occurrences. Then

$$\forall x - - - - x **** \qquad \textit{35.2}$$

means

Everything has the property that – – – – it ****. ***35.3***

'∀*x*' is called a *universal quantifier*, and is pronounced 'for all *x*'. The same applies if '*x*' is replaced throughout by another individual variable; for example

$$\forall z - - - - z **** \qquad \textit{35.4}$$

means just the same as (35.2).

Thus the sentence

∀*x x* is identical with *x*. ***35.5***

means

Everything has the property that it is identical with it. ***35.6***

or in more idiomatic English

Everything is identical with itself. ***35.7***

Here are some other ways in which English expresses the sense of the universal quantifier; note how it combines with '→':

Every cloud has a silver lining. ***35.8***
∀*x* if *x* is a cloud then *x* has a silver lining.
∀*x* [*x* is a cloud → *x* has a silver lining]

All the bells in heaven shall ring. ***35.9***
∀*x* [*x* is a bell in heaven → *x* shall ring]

Each student must hand in homework. ***35.10***
∀*x* [*x* is a student → *x* must hand in homework]

Nobody knows the trouble I seen. *35.11*
$\forall y$ [y is a person → y doesn't know the trouble I seen]

Roses are red. *35.12*
$\forall z$ [z is a rose → z is red]

In the next two examples, we quantify a 2-place predicate:

Nothing will upset x. *35.13*
$\forall y$ y won't upset x.

y doesn't know **anybody** who can sign his application for a passport. *35.14*
$\forall x$ [y knows x → x can't sign y's application for a passport]

Note that in (35.13), the variable chosen for the quantifier must be different from 'x'; otherwise we should have

$\forall x$ x won't upset x. *35.15*

which means 'Nothing will upset itself.' For the same reason, the variable chosen for the quantifier in (35.14) must be different from 'y'.

In (35.2), the variable 'x' is no longer serving to mark the place where a designator can be put; it is now simply part of a complex expression which means (35.3). In the terminology of p. 66, this implies that the occurrences of 'x' in (35.2) are not free. Instead we say that the occurrences of 'x' in (35.2) are *bound*. Likewise the two occurrences of 'y' in

$\forall y$ y won't upset x. *35.16*

are bound. However, the occurrence of 'x' in (35.16) is free, since it was not involved in the quantification. (35.16) is therefore a 1-place predicate, formed by quantifying the 2-place predicate

y won't upset x. *35.17*

In (35.15), all three occurrences of 'x' are bound, and there are no other individual variables; (35.15) is a declarative sentence.

(ii) *Existential quantification*

Suppose we have a predicate, written schematically as

– – – – x ****, *35.18*

in which the individual variable 'x' has one or more free occurrences. Then

$\exists x$ – – – – x **** **35.19**

means

At least one thing has the property that – – – – it ****. **35.20**

'$\exists x$' is called an *existential quantifier*, and is pronounced 'there is x such that'. As with universal quantification, the same definition applies if 'x' in (35.19) is everywhere replaced by another individual variable.

For example,

$\exists x$ x has got into the tank **35.21**

means

At least one thing has the property that it has got into the tank. **35.22**

or in more idiomatic English

Something's got into the tank. **35.23**

Here are some other ways in which English expresses the sense of the existential quantifier; note how it combines with '$\wedge$':

There is a tavern in the town. **35.24**
$\exists x$ x is a tavern and x is in the town.
$\exists x$ [x is a tavern $\wedge$ x is in the town]

I heard it from **one** of your friends. **35.25**
$\exists y$ [y is a friend of yours $\wedge$ I heard it from y]

A mad dog has bitten x. **35.26**
$\exists z$ [z is a mad dog $\wedge$ z has bitten x]

Some people prefer z. **35.27**
$\exists x$ [x is a person $\wedge$ x prefers z]

The translation in (35.27) is not perfect, since **some people** often implies more than one person. But it's not too bad.

The variable chosen for the quantifier in (35.26) must be different from 'x', for the same reason as in (35.13); the sentence

$$\exists x\ [x \text{ is a mad dog} \wedge x \text{ has bitten } x] \qquad \textbf{35.28}$$

means 'Some mad dog has bitten itself.'

Just as with universal quantification, the occurrences of 'x' in (35.19) are not free, and they are said to be *bound*. Existential quantification binds variables, and thus reduces the number of places in a predicate.

Exercise 35. Express each of the following (from Shakespeare's *Macbeth*) as faithfully as possible, using a sentence or predicate which starts with a universal or existential quantifier.

1. Every noise appals me.
2. Something wicked this way comes.
3. I have a strange infirmity.
4. Their candles are all out.
5. He has no children.
6. Murders have been performed.
7. x is a tale told by an idiot.
8. None of woman born shall harm x.

In the terminology of section 34, the universal quantifier has a profile in the second group, while the existential quantifier has a profile in the first. What about domain of quantification?

We leave the answer open: the domain can be whatever is convenient for the purpose in hand. It's usually convenient to stipulate the same domain of quantification for all the quantifiers which occur in the set of sentences under discussion; for technical simplicity we shall expect the domain of quantification to include the primary reference of any designator which occurs in the set of sentences. In section 40 we shall see that many logicians require the domain of quantification to have at least one thing in it.

36. Quantifier Rules

A sentence in which no universal or existential quantifiers occur is said to be *quantifier-free*. In this section we shall see that the inconsistency of a set X of sentences can often be demonstrated by proving the inconsistency of certain other sentences which are quantifier-free. These other sentences are called *Herbrand sentences* of the set X (after Jacques Herbrand, who suggested this approach in 1930).

The Herbrand sentences of the set X are constructed from X by certain rules, which we shall describe in a moment. But here is an example of what to expect. Suppose X is the set

$\forall x$ *x* doesn't excite me. Gertie excites me. ***36.1***

Then we shall find that X has three Herbrand sentences, which are

Gertie doesn't excite me. I don't excite myself. ***36.2***
Gertie excites me.

(36.2) is blatantly inconsistent; we infer that X is inconsistent too.

The Herbrand sentences of a set X come from three sources. First, *every quantifier-free sentence in* X *is a Herbrand sentence of* X. The second and third kinds of Herbrand sentence are found by dropping universal and existential quantifiers respectively; this is to be done in ways which are justified by two rules, the $\forall x\phi$ Rule and the $\exists x\phi$ Rule, which we shall now describe.

(i) *The* $\forall x\phi$ *Rule*

Intuitively, this rule asserts that if everything has some property, then any one named thing has the property. For example, if everyone has a skeleton in his or her cupboard, then even the well-known public benefactor Sir Jasper Virtue has a skeleton in his cupboard. More precisely stated, the rule is as follows:

Suppose X *is a set of declarative sentences, among which is a sentence of the form '*$\forall x\phi$*', and suppose* D *is a designator which occurs in a sentence in* X. *Suppose that* ψ *is the sentence got by putting* D *in place of every free occurrence of 'x' in* ϕ. *Then*

X. *Therefore* ψ.

is a valid argument.

The variable 'x' in this rule should be understood as a typical individual variable; the rule holds if 'x' is replaced by any other individual variable throughout.

For example, suppose X is the set

$\forall z$ [z is an element $\rightarrow$ z has an atomic number]. ***36.3***
Rhenium is an element.

We can take ϕ to be

[z is an element $\rightarrow$ z has an atomic number] ***36.4***

and D to be 'Rhenium', so that ψ is

[Rhenium is an element $\rightarrow$ Rhenium has an atomic number]. ***36.5***

The $\forall x\phi$ Rule then tells us that (36.3) entails (36.5).

The rule is justified by our First and Second Assumptions in section 28. Suppose S is a situation in which all of X is true. If D occurs in a sentence in X, then by the Second Assumption its occurrence is purely referential, so by the First Assumption it has a primary reference in S; this primary reference is assumed to lie in the domain of quantification, by p. 168. By the Second Assumption the occurrences of D in ψ are purely referential, so that ψ does express that ϕ is satisfied by the primary reference of D. Since '$\forall x\phi$' is true in S, ψ must also be true in S.

Exercise 36A. What sentences does the $\forall x\phi$ Rule allow us to deduce from the following set?

$\forall y$ [y is an older boy than John $\rightarrow$ y likes to sing].
The boy in the corner is tone-deaf.

Now suppose X and ψ are as stated in the $\forall x\phi$ Rule, and ψ is quantifier-free; *then ψ will be a Herbrand sentence of* X. With a refinement to be added in a moment, this is our second source of Herbrand sentences.

We can see that all the Herbrand sentences of X described so far are actually entailed by X, so that if these Herbrand sentences are inconsistent, X itself must certainly be inconsistent. So we can prove the inconsistency of X by producing an inconsistent set of Herbrand sentences of X, as in (36.2).

Here is another example. We consider the set of sentences

$\forall x$ [x is a pygmy chimpanzee → x can't use language]. ***36.6***
[Kanzi is a pygmy chimpanzee ∧ Kanzi can use language].

To show that (36.6) is inconsistent, it's enough to write down two of its Herbrand sentences:

[Kanzi is a pygmy chimpanzee → Kanzi can't use language]. ***36.7***
[Kanzi is a pygmy chimpanzee ∧ Kanzi can use language].

(36.7) is obviously inconsistent. (Check this if necessary, by a sentence tableau.) Thus (36.6) is proved *inconsistent.*

In proving inconsistency by this method, it is not always necessary to write down every single one of the Herbrand sentences. Thus in (36.2) we could have omitted 'I don't excite myself.'

Exercise 36B. Prove the inconsistency of each of the following sets of sentences by writing down an inconsistent set of Herbrand sentences; where necessary, check the inconsistency of the Herbrand sentences by a tableau.

1. $\forall x$ [x is odd ∨ x is even].
   The number ½ is neither odd nor even.
2. $\forall x$ [x is a person → x's I.Q. never varies by more than 15].
   Case 946 is a girl whose I.Q. has varied between 142 and 87.
3. $\forall x$ [x is Austrian → x is of Alpine type].
   $\forall x$ [x is of Alpine type → x has a broad head].
   Horst is Austrian, but doesn't have a broad head.
4. $\forall y$ [[y is a cat ∧ y has two ginger parents] → y is ginger].
   $\forall x$ [x is a female cat → x is not ginger].
   The female cat over the road has two ginger parents.

We get still more Herbrand sentences of X to be produced by applying the $\forall x\phi$ Rule to X *together with any Herbrand sentences already constructed.* For example, let X be the set of sentences

$\forall x$ [x is a widow $\rightarrow$ the spouse of x is dead]. **36.8**
$\forall x$ x is not dead.
Maxine is a widow.

As a first step, we can write down two Herbrand sentences of X:

Maxine is a widow. **36.9**
[Maxine is a widow $\rightarrow$ the spouse of Maxine is dead].

Now (36.9) presents us with a new designator: 'the spouse of Maxine'. Applying the $\forall x\phi$ Rule with this designator to the second sentence of (36.8), we can add the further Herbrand sentence of X:

The spouse of Maxine is not dead. **36.10**

(36.9) and (36.10) together are inconsistent, proving that (36.8) is *inconsistent.*

Exercise 36C. By repeating the process just described as many times as we like, we can produce infinitely many different Herbrand sentences of (36.8). Write down three more besides those in (36.9) and (36.10). (Use the first sentence of (36.8) again with the new designator.)

(ii) *The $\exists x\phi$ Rule*

Intuitively, this rule says that one can never create an inconsistency by naming a thing, provided that the name is not already in use for something else. More precisely:

Suppose X *is a set of declarative sentences, among which is a sentence of the form '$\exists x\phi$' and suppose* D *is a proper name which occurs nowhere in* X. *Suppose that ψ is the sentence got by putting* D *in place of every free occurrence of 'x' in ϕ. Then ψ can be added to* X *without creating an inconsistency.*

As with the $\forall x\phi$ Rule, 'x' can be replaced throughout by any other individual variable. The rule is justified much as the $\forall x\phi$ Rule; if it's true in situation S that something satisfies ϕ, then there is a situation just like S except that some named thing satisfies ϕ.

It's important that the proper name should be a new one, lest we baptize two different objects with the same name. A large supply of unused proper names would be convenient. You can use your imagination for this. But the custom is to take as proper names the letters 'b', 'c', 'd', 'b_1', 'b_2', etc.; these letters are known as *individual constants*.

For example, suppose X is the set

$\forall x$ [x is an anarchist $\rightarrow$ x belongs to the Left]. ***36.11***
$\forall x$ [x belongs to the Left $\rightarrow$ x favours state control].
$\exists x$ [x is an anarchist $\wedge$ x doesn't favour state control].

The $\exists x\phi$ Rule tells us that if X is consistent, then so is the set Y consisting of X together with the sentence

[b is an anarchist $\wedge$ b doesn't favour state control]. ***36.12***

In fact, of course, Y is inconsistent. We can see this by writing down three of its Herbrand sentences, which are visibly inconsistent:

[b is an anarchist $\wedge$ b doesn't favour state control]. ***36.13***
[b is an anarchist $\rightarrow$ b belongs to the Left].
[b belongs to the Left $\rightarrow$ b favours state control].

Since Y is inconsistent, we must conclude that X was already inconsistent. Thus is (36.11) proved *inconsistent.*

If X and ψ are as stated in the $\exists x\phi$ Rule, and ψ is quantifier-free, *then* ψ *will be a Herbrand sentence of* X. This forms our third and final source of Herbrand sentences.

Thus (36.12) was a Herbrand sentence of (36.11), and hence so were (36.13). Note that in finding the Herbrand sentences (36.13) we used the $\exists x\phi$ Rule first, and then the $\forall x\phi$ Rule with the new proper name; this is usually the best strategy when one is faced with both universal and existential quantifiers.

Exercise 36D. Prove the inconsistency of each of the following sets of sentences by writing down an inconsistent set of Herbrand sentences. (Satisfy yourself that the set you write down is inconsistent.)

1. $\forall x$ [x is female $\rightarrow$ x is not a jockey].
   $\exists x$ [x is a jockey $\wedge$ x is female].
2. $\forall x$ [x does the cleaning $\rightarrow$ x doesn't go out to work].
   $\forall x$ [x is male $\rightarrow$ x goes out to work].
   $\exists x$ [x is male $\wedge$ x does the cleaning].

3. $\forall x$ [x is male $\vee$ x is female].
   $\forall x$ [x is male $\rightarrow$ x has just forty-six chromosomes].
   $\forall x$ [x is female $\rightarrow$ x has just forty-six chromosomes].
   $\exists x$ x has forty-seven chromosomes.
4. $\forall x$ [x is male $\rightarrow$ the mate of x is female].
   $\forall x$ [x is female $\rightarrow$ x is not male].
   $\exists$ [x is male $\wedge$ the mate of x is male].

Herbrand sentences can also be used for showing the validity of arguments, once we can deal with a '$\neg$' at the beginning of a sentence. Fortunately this is easy, by virtue of the following two rules:

(iii) *The* $\neg\forall x\phi$ *Rule*: '$\neg\forall x\phi$' *is true in exactly the same situations as* '$\exists x\neg\phi$'.

(iv) *The* $\neg\exists x\phi$ *Rule*: '$\neg\exists x\phi$' *is true in exactly the same situations as* '$\forall x\neg\phi$'.

For example,

> Not all has been lost. **_36.14_**
> $\neg\forall x$ x has been lost.

is true in precisely the same circumstances as

> Something has not been lost. **_36.15_**
> $\exists x\neg$ x has been lost.

We shall illustrate the use of these rules by proving the validity of the following argument:

> Bank-notes all carry a metal strip. Anything with a metal strip can be detected by X-rays. **_36.16_**
> *Therefore* bank-notes can be detected by X-rays.

The first step is to translate into quantifier notation, as in section 35:

> $\forall x$ [x is a bank-note $\rightarrow$ x has a metal strip]. **_36.17_**
> $\forall x$ [x has a metal strip $\rightarrow$ x can be detected by X-rays].
> *Therefore* $\forall x$ [x is a bank-note $\rightarrow$ x can be detected by X-rays].

When we form the counterexample set of (36.17), '*Therefore*' is replaced by '¬'. Then we can use the $\neg\forall x\phi$ Rule to replace '$\neg\forall x$' by '$\exists x\neg$', and paraphrase the counterexample set as

$\forall x$ [x is a bank-note → x has a metal strip]. ***36.18***
$\forall x$ [x has a metal strip → x can be detected by X-rays].
$\exists x$ ¬ [x is a bank-note → x can be detected by X-rays].

(36.18) has the following Herbrand sentences, which are inconsistent (as a sentence tableau would show):

¬ [b is a bank-note → b can be detected by X-rays]. ***36.19***
[b is a bank-note → b has a metal strip].
[b has a metal strip → b can be detected by X-rays].

Thus (36.16) is proved *valid.*

Exercise 36E. Prove the validity of each of the following arguments, by writing down inconsistent sets of Herbrand sentences. (Satisfy yourself that the sets are inconsistent.)

1. No ground-feeding birds are brightly coloured. Dunnocks are ground-feeding birds. *Therefore* no dunnocks are brightly coloured.
2. Some finches crack cherry seeds. All finches are birds. All birds which crack cherry seeds have massive beaks. *Therefore* some finches have massive beaks.
3. All the birds are either chiff-chaffs or willow warblers. The birds are singing near the ground. Chiff-chaffs don't sing near the ground. *Therefore* the birds are all willow warblers.
4. Trumpeter bullfinches can sing two notes at once. Trumpeter bullfinches are birds. There are some trumpeter bullfinches. *Therefore* there are some birds that can sing two notes at once.

Predicate Logic

In the next few sections we shall amalgamate nearly everything we have done so far. The result is *first-order predicate logic*, the crowning achievement of modern logic. The first systematic description of this form of logic was published by two mathematicians, David Hilbert and Wilhelm Ackermann, in 1928. Eight years later Gerhard Gentzen invented several kinds of proof calculus for first-order predicate logic. One was the direct ancestor of the tableau calculus. Another (discovered also by Stanisław Jaśkowski) was the *natural deduction calculus* which you can find in many textbooks.

37. Logical Scope

When we first encountered the logical analyst, in sections 16–19, she was busy translating English sentences into truth-functor notation. On our second visit we find that she knows about quantifiers too.

Translating into quantifier notation is a fairly subtle business; unlike truth-functor analyses, it nearly always calls for a major reorganization of the sentence. According to section 35, a sentence or predicate can have '∀' or '∃' put at its head if it can be paraphrased

Everything has the property that . . . ***37.1***

or

At least one thing has the property that . . . ***37.2***

Neither (37.1) nor (37.2) is exactly a common turn of phrase in English, so we must expect wholesale paraphrasing.

If a sentence or predicate can be paraphrased as (37.1), we say it has *overall form* ∀. If it can be paraphrased as (37.2), we say it has *overall form* ∃. Analysis of complex sentences proceeds just as in section 19: we look for the overall form of the whole sentence, and then for the overall forms of the parts inside it, moving from larger phrases to smaller.

For example, suppose we wish to analyse

Brand X doesn't remove all kinds of stain. ***37.3***

The overall form of (37.3) is 'It's not true that ϕ'; so we first put

¬ Brand X removes all kinds of stain. ***37.4***

The part after the '¬' sign in (37.4) has overall form ∀, and is easily rendered as

∀y [y is a kind of stain → Brand X removes y]. ***37.5***

Adding the '¬', we have the final analysis:

¬∀y [y is a kind of stain → Brand X removes y]. ***37.6***

For our next example, we analyse

Brand X doesn't remove any kinds of stain. ***37.7***

(37.7) has overall form ∀, and we can begin by paraphrasing it as

∀y [y is a kind of stain → Brand X doesn't remove y]. ***37.8***

The phrase 'Brand X doesn't remove y' has overall form '¬ ϕ', and can be translated into

¬ Brand X removes y. ***37.9***

Incorporating this back into (37.8), we have the final paraphrase

∀y [y is a kind of stain → ¬ Brand X removes y]. ***37.10***

The two English sentences which we have just analysed seem very similar at first glance, yet their analyses are quite different. Why? Let us contrast the phrase-markers of the two analyses. That for (37.6) is:

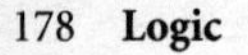

37.11

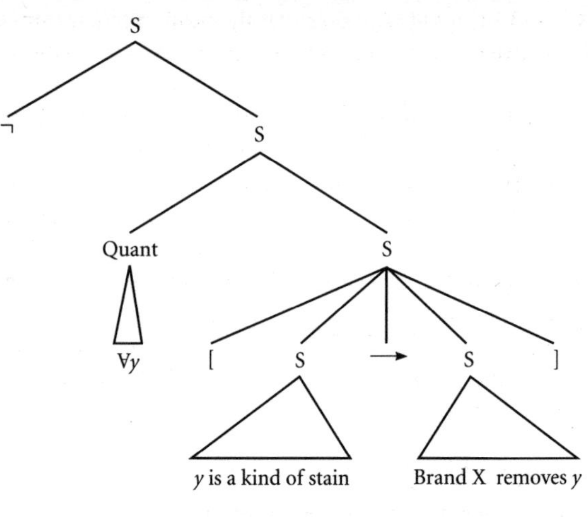

The phrase-marker for (37.10) is:

37.12

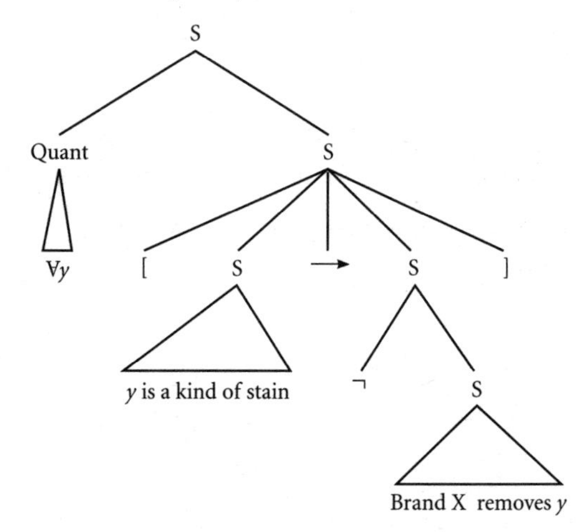

In (37.11) the scope of '¬' is the whole sentence, and includes that of '∀*y*'; but in (37.12) the scope of '∀*y*' includes that of '¬'. From a purely grammatical point of view this is surprising, because the two English sentences (37.3) and (37.7) must have very much the same phrase-markers as each other, and there is no suggestion that 'all' and 'any' have different scopes. It seems that the grammatical scope of a word such as 'any' is a very poor guide to the symbolic translation.

However, (37.3) and (37.7) do certainly mean different things and the difference between the two phrase-markers above records this difference of meaning.

Logicians often talk of the *logical scope* of an occurrence of **any** (or similar words), meaning the scope of a symbolic quantifier that would be used to translate it. Thus a logician might say that the logical scope of 'any' in (37.7) is the whole sentence while the logical scope of 'all' in (37.3) doesn't include the negation. The logical scope of a word is an important guide to the meaning of the sentence.

In fact we have just uncovered an important difference between **all** and **any**: **any** *tends to have larger logical scope than* **all**. Here is another pair of examples to illustrate this:

If **any** of the brakes hold, the train will halt. ***37.13***
$\forall x$ [[x is a brake $\wedge$ x will hold] $\rightarrow$ the train will halt].

If **all** of the brakes hold, the train will halt. ***37.14***
[$\forall x$ [x is a brake $\wedge$ x will hold] $\rightarrow$ the train will halt].

Every is like **all**; it tends to have a small logical scope, as in the following examples:

I don't know **anything**. ***37.15***
$\forall x \neg$ I know x.

I don't know **everything**. ***37.16***
$\neg \forall x$ I know x.

Exercise 37A. Analyse with the aid of '$\forall$' or '$\exists$':

1. The room isn't heated at all times.
2. The room is unheated at all times.
3. The room isn't heated at any time.
4. There are times at which the room isn't heated.
5. There aren't any times at which the room is heated.

So far, we have not considered any sentences in which two quantifiers occur. We shall remedy this at once, by analysing

Some girl won all the prizes. ***37.17***

This has overall form $\exists$, and we can start by writing

$$\exists x\ [x \text{ is a girl} \wedge x \text{ won all the prizes}]. \qquad \textbf{\textit{37.18}}$$

Then 'x won all the prizes' has overall form $\forall$:

$$\forall y\ [y \text{ was a prize} \rightarrow x \text{ won } y]. \qquad \textbf{\textit{37.19}}$$

Fitting (37.19) back into (37.18), we have the final analysis

$$\exists x\ [x \text{ is a girl} \wedge \forall y\ [y \text{ was a prize} \rightarrow x \text{ won } y]]. \qquad \textbf{\textit{37.20}}$$

Another example:

Each of the prizes was won by a girl. ***37.21***

Here the overall form is $\forall$:

$$\forall y\ [y \text{ was a prize} \rightarrow y \text{ was won by a girl}]. \qquad \textbf{\textit{37.22}}$$

The second phrase inside (37.22) has overall form $\exists$:

y was won by a girl ***37.23***
$\exists x\ [x \text{ is a girl} \wedge x \text{ won } y]$

so that the final analysis is:

$$\forall y\ [y \text{ was a prize} \rightarrow \exists x\ [x \text{ is a girl} \wedge x \text{ won } y]]. \qquad \textbf{\textit{37.24}}$$

Contrast (37.17) with (37.21). In (37.17) a single girl takes all, while (37.21) allows each prize to have been won by a different girl. The quantifier analyses (37.20) and (37.24) bring out the precise difference: they show that in (37.17) 'some' has greater logical scope than 'all', whereas in (37.21), 'each' has greater logical scope than 'a'.

In English, logical scopes are determined in a complicated and obscure way. It has something to do with the exact choice of words (**some/a; all/each**) and something to do with the order of the words in the sentence. The rules are so tangled that different people often find they interpret one and the same sentence in quite different ways. For this reason, you should regard the next exercise as a piece of research into your own brand of English; there are no universal right answers.

Exercise 37B. In each of the following sentences, which of the two bold words has the greater logical scope?

1. **A** girl won **all** the prizes.
2. **All** the prizes were won by **some** girl.
3. **A** girl won **every** prize.
4. **Every** prize was won by **some** girl.
5. **Every** one of the prizes was won by **a** girl.
6. **Some** girl won **each** prize.

When Phineas Barnum fooled all the people some of the time, did he fool everybody at once, or did he merely fool each person at some time or other?

Apparently no natural language is altogether happy with complicated arrays of quantifiers. It is true that in favourable cases, English can handle a fair amount of quantificational complexity:

> Each of us must admit that there have been times in our lives when we have felt that everyone round us has posed some kind of threat to all our values. *37.25*

But some other assertions, which are theoretically no more complex than (37.25), become almost unintelligible when they are set out in brute English:

> For every person and every age, and every positive number, there is a second positive number such that at any age which differs from the first-mentioned age by fewer days than the latter positive number, the person's height differs from his or her height at the first-mentioned age by less than the former positive number of inches. *37.26*

Mathematicians will recognize (37.26) as the statement that a person's height is a continuous function of his or her age. Slight changes in logical scope would lead to the quite different statements that a person's height was a continuous function of his age uniformly with respect to people, or with respect to age. These are examples of vitally important mathematical notions. It should be no surprise that universal and existential quantifiers were first discovered by two mathematician-philosophers

(Gottlob Frege 1879, C. S. Peirce 1883) at just the time when the notions of continuity and uniform continuity were being absorbed into the corpus of mathematics.

When two or more quantifiers occur in a sentence, we must be careful about our choice of individual variables. We shall make it a rule that *no quantifier should occur within the scope of another occurrence of a quantifier with the same variable.*

For example, we analyse:

Every house has a freezer and a waste disposal unit. ***37.27***

The first steps are clear:

$\forall x$ [x is a house $\rightarrow$ [x has a freezer $\wedge$ x has a waste disposal unit]]. ***37.28***

The variable of the quantifier used in 'x has a freezer' must be different from 'x', because this phrase lies within the scope of '$\forall x$' at the front; we can choose 'y':

$\forall x$ [x is a house $\rightarrow$ [$\exists y$ [y is a freezer $\wedge$ x has y] $\wedge$ x has a waste disposal unit]] ***37.29***

Now 'x has a waste disposal unit' lies within the scope of '$\forall x$', but not within that of '$\exists y$'. So we must avoid adding a quantifier with 'x', but there is no objection to another 'y':

$\forall x$ [x is a house $\rightarrow$ [$\exists y$ [y is a freezer $\wedge$ x has y] $\wedge$ $\exists y$ [y is a waste disposal unit $\wedge$ x has y]]]. ***37.30***

Exercise 37C. Express each of the following sentences as faithfully as possible, using truth-functor symbols and universal and existential quantifiers.

1. If there are any starlings nesting here, then I'll shoot them.
2. If there are any starlings nesting here, then that bird is a starling.
3. That bird has a longer bill than any finch.
4. That bird has a bill no longer than some finches.
5. That bird has a bill no longer than any starling.
6. Not all starlings have longer bills than any finch.
7. There are finches with longer bills than any starling.
8. For any finch, there is a starling with a longer bill.

9. All the birds nesting here, except possibly finches, have long bills.
10. Among the birds nesting here, only the starlings have long bills.

The notion of logical scope is not always wholly determinate, because it may be possible to analyse one English sentence in two quite different ways, both equally correct. Exercise 37C.2 is an example; see the answers. Even the logician's tools may break if you lean too hard on them.

38. Analyses Using Identity

In section 35 we chose to adopt symbols for just two types of quantifier, the universal and the existential. Why not for other kinds of phrase too, such as **at least two** or **most**? Part of the answer is that many of these other quantifier phrases can be paraphrased by means of the two quantifiers we already have, together with '='.

(i) At least, at most *and* exactly

We can express the senses of **at least one, at least two** and so on, by combining '∃' with '=':

I know **at least one** pop-star. ***38.1***
$\exists x\,[x \text{ is a pop-star} \wedge \text{I know } x]$

I know **at least two** pop-stars. ***38.2***
$\exists x_1 \exists x_2\,[\neg\, x_1 = x_2 \wedge [[x_1 \text{ is a pop-star} \wedge \text{I know } x_1] \wedge [x_2 \text{ is a pop-star} \wedge \text{I know } x_2]]]$

I know **at least three** pop-stars. ***38.3***
$\exists x_1 \exists x_2 \exists x_3\,[[[\neg\, x_1 = x_2 \wedge \neg\, x_1 = x_3] \wedge \neg\, x_2 = x_3] \wedge [[[x_1 \text{ is a pop-star} \wedge \text{I know } x_1] \wedge [x_2 \text{ is a pop-star} \wedge \text{I know } x_2]] \wedge [x_3 \text{ is a pop-star} \wedge \text{I know } x_3]]]$.

The exact arrangement of the brackets is not important.

Then **at most one, at most two** and so on can be expressed by saying 'not at least two', 'not at least three', etc.:

I know **at most one** prince. ***38.4***
$\neg$ I know at least two princes.

I know **at most two** princes. ***38.5***
$\neg$ I know at least three princes.

Exactly means 'at most and at least':

I know **exactly one** name-dropper. ***38.6***
[I know at least one name-dropper $\wedge$ I know at most one name-dropper].

Written out in full, the paraphrase of (38.6) is quite lengthy. There is a neater way of expressing **exactly one**:

I know **exactly one** name-dropper. ***38.7***
$\exists x \forall y\ [x = y \leftrightarrow [y \text{ is a name-dropper} \wedge \text{I know } y]]$.

Exercise 38A. Write out paraphrases of the following, using '$\forall$' '$\exists$' and '$=$'.

1. There are at least two mistakes.
2. There are at least four mistakes.
3. More than one person has pointed out the mistakes.
4. There are precisely two hemispheres.
5. Only Sir Henry is allowed to use that bath.

(ii) **The**

A sentence of the form

The S is a P. ***38.8***

is true when there is precisely one S under discussion, and every S under discussion is a P. This suggests the paraphrase

There is exactly one S, and every S is a P. ***38.9***

As it stands, the paraphrase (38.9) is clearly wrong; for example the following two sentences obviously mean quite different things:

The boy is a genius. ***38.10***

There is exactly one boy, and every boy is a genius. ***38.11***

But the reason why (38.11) fails to paraphrase (38.10) is simply that the domain of quantification demanded by (38.11) is far too large. As we noted on p. 163, **every** allows a large domain of quantification, while **the** is used to narrow the domain to those things which are immediately in hand. If we use '∀', '∃' and '=' to symbolize (38.11), and then stipulate a small enough domain of quantification, the resulting paraphrase of (38.10) will be quite adequate.

With this caution on domains of quantification, we therefore paraphrase

The boy is a genius. ***38.12***
[$\exists x \forall y$ [$x = y \leftrightarrow y$ is a boy] $\wedge$ $\forall x$ [x is a boy $\rightarrow$ x is a genius]].

The same style can be used to eliminate other definite descriptions:

Ahmed loves **his wife** dearly. ***38.13***
[$\exists x \forall y$ [$x = y \leftrightarrow y$ is a wife of Ahmed] $\wedge$ $\forall x$ [x is a wife of Ahmed $\rightarrow$ Ahmed loves x dearly]].

Sentences of the form 'A is the B' can be paraphrased more simply:

Today is **the day of liberation.** ***38.14***
$\forall y$ [today $= y \leftrightarrow y$ is a day of liberation].

There are also special styles of analysis which can be used for superlatives such as **the biggest, the most embarrassing**, and so forth:

Harrow United is **the best team.** ***38.15***
Harrow United is a team and is better than any other team.
[Harrow United is a team $\wedge$ $\forall y$ [[y is a team $\wedge$ $\neg y =$ Harrow United] $\rightarrow$ Harrow United is better than y]].

Chicago has **the worst crime-rate.** ***38.16***
$\forall y$ [$\neg y =$ Chicago $\rightarrow$ Chicago has a worse crime-rate than y].

Occasionally, definite descriptions are used in a way which is not meant to imply that just one thing fits the description; this occurs most often after **is** and **was.** In such cases the paraphrases suggested above should not be used. Thus:

Bob is **the proud owner of a Rolls Royce.** ***38.17***
NOT: $\forall y$ [Bob $= y \leftrightarrow y$ is a proud owner of a Rolls Royce].

(There may be other proud owners of Rolls Royces.) Non-count nouns (see p. 122) lead to definite descriptions which don't admit the question

'How many?'; here again our paraphrases are not appropriate. For example:

The music of Schubert is more popular than ever. ***38.18***
NOT: [$\exists x \forall y$ [$x = y \leftrightarrow y$ is a music of Schubert] $\wedge$ $\forall x$ [x is a music of Schubert $\rightarrow$ x is more popular than ever]].

Naturally you will also avoid using the above paraphrases where an occurrence of a definite description is not purely referential:

Disraeli became **the Prime Minister.** ***38.19***
NOT: [$\exists x \forall y$ [$x = y \leftrightarrow y$ is Prime Minister] $\wedge$
$\forall x$ [x is Prime Minister $\rightarrow$ Disraeli became x]].

Exercise 38B. Where possible, eliminate the bold definite descriptions in the following sentences by paraphrasing as above. Where it's impossible, say so.

1. **The book** is bound in vellum.
2. We've avoided **her stern.**
3. Joseph was **the son of Jacob.**
4. **The woman in blue** is my mother.
5. I hope to be **the first to congratulate you.**
6. The shepherds were seated on **the ground.**
7. Cassius is **the greatest.**

The paraphrases above raise some severe problems of scope when we try to apply them to complex sentences. These problems are not of our making – they reflect the great subtlety of the use of definite descriptions in English. For example, compare

I'm not **the man you lent £10 to.** ***38.20***
I'm not **the tallest man in the room.** ***38.21***

(38.20) would normally be understood to imply that there is some one man you lent £10 to; it could be paraphrased in the style above, as

[$\exists x \forall y$ [$x = y \leftrightarrow y$ is a man you lent £10 to] $\wedge$ $\forall x$ [x is a man you lent £10 to $\rightarrow \neg\, x =$ me]]. ***38.22***

But (38.21) doesn't carry a similar implication; it could well be true because there is no tallest man in the room – say, if there are three men of the same height, all taller than anybody else in the room. The correct paraphrase is

> It's not true that I'm the tallest man in the room. ***38.23***
> $\neg$ [I'm in the room $\wedge$ $\forall y$ [[y is a man in the room $\wedge$ $\neg y$ = me] $\rightarrow$ I'm taller than y]].

The logical scope of 'the' includes that of 'not' in (38.20), but not in (38.21).

In section 34 we grouped English quantifications into four main groups. Most quantifications in groups (1), (2) and (4) can be handled by '$\forall$', '$\exists$' and '='. Not so for those in group (3); this is one of the failings of first-order predicate logic. There are plenty of logical connections between group (3) phrases, if one wanted to construct a theory around them. For example, here is an inconsistent set of sentences:

> More than half the people in the room are women. ***38.24***
> More than half the women in the room have dark hair.
> Less than a quarter of the people in the room have dark hair.

39. Predicate Interpretations

As in section 21, we wish to abbreviate. Our abbreviation schemes will be called *interpretations*, or *predicate interpretations* when we want to distinguish them from the earlier sort. These interpretations must cater for designators and predicates as well as sentences. As before, a sentence or predicate which is completely translated into symbols is called a *formula*.

An example will show what is needed. Here is a predicate interpretation, written as we shall write them.

> J : Sauvignon is the juiciest of the Bordeaux grapes. ***39.1***
> Gx : x is a species of grape.
> Mxy : x is made out of y.
> Sxy : y is sweeter than x.
> b : Blue Burgundy
> c : Pinot Noir
> d : Grenache

Capital letters (such as '*J*') have declarative sentences assigned to them. A capital letter followed by one or more individual variables without repetitions (such as '*Gx*', '*Mxy*' or '*Sxy*') has assigned to it a predicate in which just the same variables have free occurrences. An individual constant (such as '*b*', '*c*', '*d*') has a designator assigned to it.

Here are some sample abbreviations by the interpretation (39.1):

Sauvignon is the juiciest of the Bordeaux grapes. ***39.2***
J

Pinot Noir is a species of grape. ***39.3***
Gc

Blue Burgundy is made out of Pinot Noir. ***39.4***
Mbc

Grenache is sweeter than *y*. ***39.5***
Syd

Note the order of the symbols in (39.5), compared with:

y is sweeter than Grenache. ***39.6***
Sdy

More complex sentences can also be abbreviated. For example, the sentence

There is a species of grape which is sweeter than Pinot Noir. ***39.7***

can first be analysed as

$\exists x$ [x is a species of grape $\wedge$ x is sweeter than Pinot Noir]; ***39.8***

then the interpretation (39.1) can be used to abbreviate this to the formula

$\exists x\ [Gx \wedge Scx]$. ***39.9***

A little paraphrasing will help us to symbolize more sentences:

Grenache is at least as sweet as Pinot Noir. ***39.10***
Pinot Noir is not sweeter than Grenache.
$\neg\ Sdc$

Exercise 39A. Using the following interpretation:

Ex : x is an epic.
Sxy : x is shorter than y.
Wxy : x wrote y.
b : Beowulf
c : the Odyssey
d : Homer

translate each of the following into a symbolic formula:

1. The Odyssey is an epic.
2. Homer didn't write Beowulf.
3. The Odyssey and Beowulf are not the same length.
4. Homer wrote an epic which is longer than Beowulf.
5. Homer didn't write just one epic.
6. Beowulf is not the shortest epic.
7. The Odyssey and Beowulf are not by the same author.
8. Any epics there may be that are shorter than Beowulf weren't written by Homer.
9. Whatever epic we consider, Homer wrote a longer one.
10. There are just two epics which are longer than the Odyssey, and Homer wrote one of them but not the other.

The classification of binary relations in section 31 can be recast in terms of formulae. For example, to say that the relation expressed by the predicate 'Rxy' in the domain of quantification is reflexive, we write

$$\forall x Rxx. \qquad \textbf{39.11}$$

Similarly, we can convey that this relation is transitive by writing

$$\forall x \forall y \forall z[[Rxy \wedge Ryz] \rightarrow Rxz]. \qquad \textbf{39.12}$$

Exercise 39B. Write down formulae which say that the relation expressed by 'Rxy' in the domain of quantification has the following properties:

1. Irreflexive.
2. Symmetric.

3. Asymmetric.
4. Intransitive.
5. Non-reflexive.
6. Connected. (This needs '='.)

---

40. Predicate Tableaux

We must reconsider tableaux. As an instrument for testing consistency, they served us faithfully up to section 25. Unfortunately there is no hope of redesigning them to test the consistency of sets of sentences with quantifiers, since Alonzo Church proved in 1936 that there cannot be a systematic method for testing consistency of such sentences. But we can still employ closed tableaux to *prove inconsistency.*

The method is an ingenious combination of sentence tableaux and Herbrand sentences. First we reinterpret the derivation rules of sentence tableaux. The derivation rules of the form

40.1

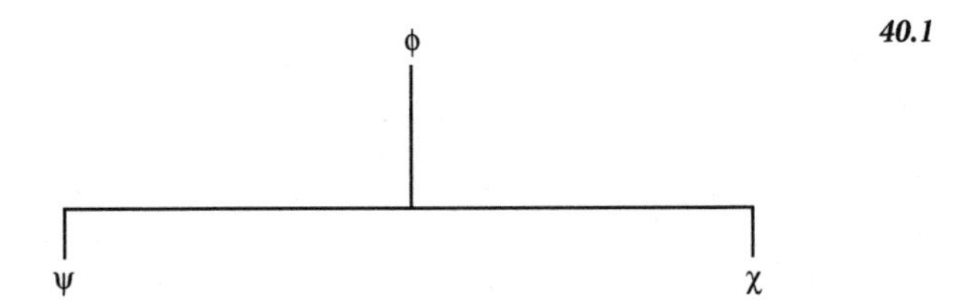

will now be interpreted as saying: *if a set of sentences containing ϕ is consistent, then either ψ can be added to the set without creating an inconsistency, or χ can.* Likewise the rules of the form

40.2

will now be read as saying: *if a set of sentences containing ϕ is consistent, then ψ and χ can be added to the set without creating an inconsistency.*

On our earlier reading (as in section 10), (40.1) would have said also that if either ψ or χ is true in a situation, then ϕ is true in it; and similarly with (40.2). Interpreted in the new way, *a closed tableau still proves that a set of sentences is inconsistent*; but an unclosed tableau proves nothing at all. Ticks, which were used in earlier sections to check that an unclosed tableau was finished, are no longer needed – we have lost interest in unclosed tableaux.

When the derivation rules are read in the new way, we can immediately add six to their number, as follows:

I. ϕ
$\mathrm{D} = \mathrm{E}$
|
ψ

II. ϕ
$\mathrm{E} = \mathrm{D}$
|
ψ

(I and II:) where the designator D occurs in ϕ, and ψ is the result of replacing one or more occurrences of D in ϕ by occurrences of E.

III. $\forall x\phi$
|
ψ

where there is a designator D which has already occurred in the branch to which ψ is added, and ψ is the result of replacing each free occurrence of 'x' in ϕ by an occurrence of D.

IV. $\exists x\phi$
|
ψ

where there is a proper name D which has not occurred anywhere in the branch to which ψ is added, and ψ is the result of replacing each free occurrence of 'x' in ϕ by an occurrence of D.

V. $\neg\forall x\phi$
|
$\exists x\neg\phi$

VI. $\neg \exists x\phi$

$|$

$\forall x \neg \phi$

Rules I and II are easily recognized as tableau forms of Leibniz's Rule from section 29. If the two upper sentences of one of these rules occur in a branch, and the branch is consistent, then the sentence ψ can certainly be added without creating an inconsistency, since the other sentences already entail it.

Rules III–VI likewise are the tableau forms of the quantifier rules which we studied in section 36. The 'x' can be replaced by any other individual variable, provided we don't confuse two variables within one rule. For example Rule V should be read as including

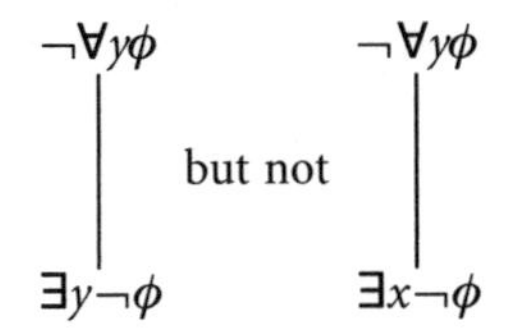

The Identity Rule from section 29 is not forgotten; it tells us that *we can close any branch in which a sentence of the form* '$\neg$ D = D' *occurs.*

Tableaux which use the rules above, together with the rules of sentence tableaux, are known as *predicate tableaux.*

Many logicians add a seventh derivation rule. It has the effect of decreeing that every domain of quantification has at least one individual in it, and so we can reasonably call it the *Nonempty Domain Rule.* As we shall see in the next section, the reasons for adopting the Nonempty Domain Rule are historical and technical; nothing that we have done so far in this book calls for it. But for convenience here it is.

VII. $\forall x\phi$ $|$ ψ

where no designators have occurred in any sentence in the branch to which ψ is added, D is a proper name and ψ is the result of replacing each free occurrence of 'x' in ϕ by an occurrence of D.

To illustrate predicate tableaux, let us recast our proof of the inconsistency of the following set, which was (36.11) in section 36:

$\forall x$ [x is an anarchist $\rightarrow$ x belongs to the Left]. ***40.3***
$\forall x$ [x belongs to the Left $\rightarrow$ x favours state control].
$\exists x$ [x is an anarchist $\wedge$ x doesn't favour state control].

The new proof is a closed tableau:

$\forall x$ [x is an anarchist $\rightarrow$ ***40.4***
x belongs to the Left]
$\forall x$ [x belongs to the Left $\rightarrow$
x favours state control]
$\exists x$ [x is an anarchist $\wedge$
x doesn't favour state control]

|

[b is an anarchist $\wedge$ (by IV)
b doesn't favour state control]

|

[b is an anarchist $\rightarrow$ (by III)
b belongs to the Left]

|

[b belongs to the Left $\rightarrow$ (by III)
b favours state control]

|

b is an anarchist.
b doesn't favour state control.

$\neg$ b is an anarchist. | b belongs to the Left.

$\neg$ b belongs to the Left. | b favours state control.

All the examples of section 36 can be recast in tableau terms. In section 36, no sentence contained more than one occurrence of a quantifier. There was therefore no need to worry about which occurrences of individual variables were free in the sentence ϕ of the $\forall x\phi$ and $\exists x\phi$ Rules – all occurrences were automatically free. Theoretically we may now have difficulties in telling the bound from the free occurrences when we apply Rules III and IV. Since section 37, we have allowed a sentence to contain several quantifiers, and some quantifier might survive to bind an 'x' after the initial '$\forall x$' or '$\exists x$' was lopped off. But in fact this problem never arises, provided we stick to the translating procedure of section 37; when '$\forall x$' or '$\exists x$' is removed from the front, all remaining occurrences of 'x' become free. (Of course if '$\forall x$' or '$\exists x$' is taken off the front, there may be some occurrences of other individual variables such as 'y' or 'z' still in the sentence. But these will all be bound, and should be left sleeping until their respective quantifiers are reached later in the tableau.)

Since tableaux are bulky, our remaining examples will be carried out in symbols, using suitable interpretations. In symbolic formulae there is no difference between designators and proper names – both are represented by individual constants. From now on, we shall omit the connecting lines in tableaux when no branching occurs; this will reduce the clutter.

Our next example is one which, but for its length, could have been handled by the method of section 36. The following argument is on loan from Sigmund Freud:

> Criticisms which stem from some psychological need of those making them don't deserve a rational answer. When people complain that psychoanalysis makes wild and arbitrary assertions about infantile sexuality, this criticism stems from certain psychological needs of these people. *Therefore* the criticism that psychoanalysis makes wild and arbitrary assertions about infantile sexuality doesn't deserve a rational answer.† **40.5**

We take as interpretation:

> Cx : x is a criticism which stems from some psychological need of those making it. **40.6**
> Rx : x deserves a rational answer.
> b : the complaint that psychoanalysis makes wild and arbitrary assertions about infantile sexuality.

Translated into formulae by (40.6), the argument (40.5) becomes

$$\forall x[Cx \rightarrow \neg Rx].\ Cb.\ \textit{Therefore}\ \neg Rb. \qquad \textbf{40.7}$$

† Sigmund Freud, *New Introductory Lectures on Psychoanalysis*, Hogarth Press, 1933; see his Lecture XXXIV. Also in Pelican Freud Library, Penguin Books, 1973; vol. 2, Lecture 34.

The tableau proof of (40.7) is very brief; it uses Rule III once:

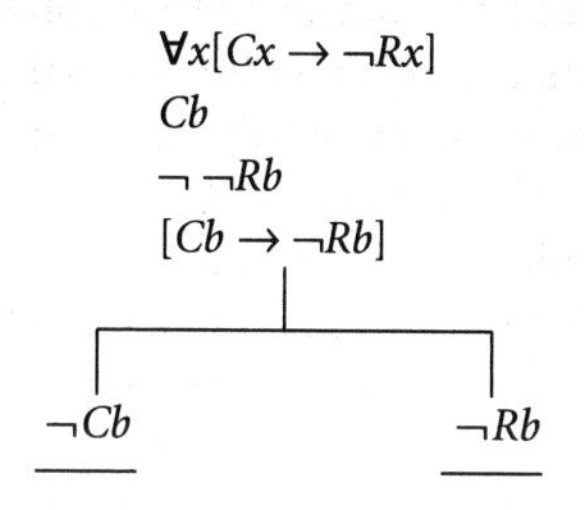

40.8

Exercise 40A. Using the interpretation given, translate the following argument (partly taken from an old American state constitution) into symbols. Prove the validity of the symbolic argument by a closed tableau.

> M : No man may be beaten with above forty stripes.
> Gx : x is a true gentleman.
> Cx : x's crime is very shamefull.
> Lx : x's course of life is vitious and profligate.
> Sx : x may be punished with shipping.
> d : John Doe Jr.

> No man may be beaten with above forty stripes; nor may any true gentleman be punished with shipping, unless his crime be very shamefull, and his course of life vitious and profligate. John Doe Jr is a true gentleman, and, though his crime is in truth very shamefull, his course of life is in no manner vitious and profligate. *Therefore* John Doe Jr may not be punished with shipping.

For our next example, we show that every irreflexive and transitive binary relation is asymmetric. In view of page 189, we can express this as follows:

$\forall x \neg Rxx.\ \forall x \forall y \forall z[[Rxy \wedge Ryz] \rightarrow Rxz].$ ***40.9***
Therefore $\forall x \forall y[Rxy \rightarrow \neg Ryx].$

The tableau runs:

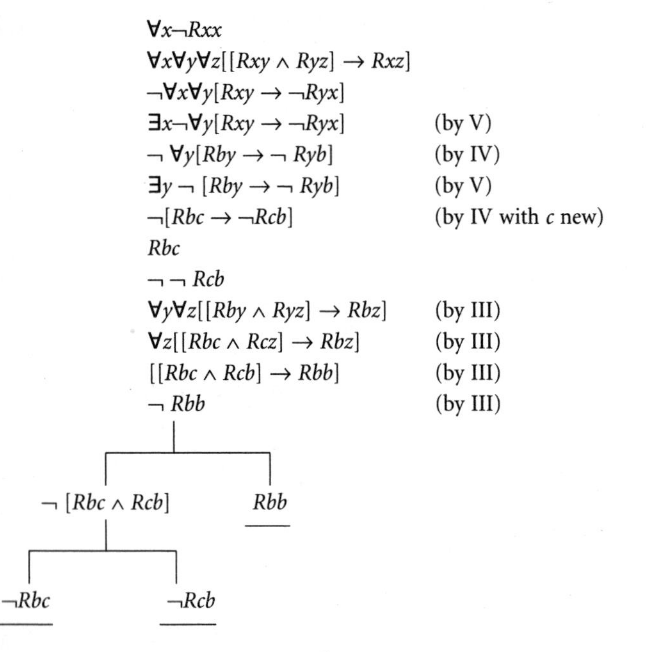

Exercise 40B. Show that every asymmetric binary relation is irreflexive. (It's easier than (40.10).)

In our next example, the placing of the quantifiers is more complicated:

> There's someone who's going to pay for all the breakages. *Therefore* each of the breakages is going to be paid for by someone. ***40.11***

For an interpretation, we take:

Px	: *x* is a person.	***40.12***
Bx	: *x* is a breakage.	
Gxy	: *x* is going to pay for *y*.	

Symbolized, (40.11) becomes:

$\exists x[Px \wedge \forall y[By \rightarrow Gxy]]$. **40.13**
Therefore $\forall y[By \rightarrow \exists x[Px \wedge Gxy]]$.

The tableau to prove (40.13) is:

40.14

$\exists x[Px \wedge \forall y[By \rightarrow Gxy]]$
$\neg\forall y[By \rightarrow \exists x[Px \wedge Gxy]]$
$\exists y \neg [By \rightarrow \exists x[Px \wedge Gxy]]$
$\neg[Bb \rightarrow \exists x[Px \wedge Gxb]]$
Bb
$\neg\exists x[Px \wedge Gxb]$
$\forall x\neg[Px \wedge Gxb]$
$[Pc \wedge \forall y[By \rightarrow Gcy]]$
Pc
$\forall y[By \rightarrow Gcy]$
$[Bb \rightarrow Gcb]$

$\neg Bb$ (closed) | Gcb
$\neg [Pc \wedge Gcb]$

$\neg Pc$ (closed) | $\neg Gcb$ (closed)

Exercise 40C. Show that any sentence which can be symbolized as follows is inconsistent:

$\exists x[Rx \wedge \forall y[Ry \rightarrow [Pxy \leftrightarrow \neg Pyy]]]$

(See Exercise 1B.4 on p. 3 for an example.)

So far, none of our examples have used the rules for identity. A well-known children's riddle will illustrate these rules:

Brother and sister have I none; **40.15**
But that man's father is my father's son.
Therefore I am that man's father.

An interpretation:

Fxy : x is the father of y. **40.16**

b : me

c : that man's father

With enough accuracy, we can symbolize (40.15) as

$\forall y[\exists x[Fxy \wedge Fxb] \rightarrow y = b]$. $\exists x[Fxc \wedge Fxb]$. **40.17**
Therefore $b = c$.

The tableau:

40.18

$$\forall y[\exists x[Fxy \wedge Fxb] \rightarrow y = b]$$
$$\exists x[Fxc \wedge Fxb]$$
$$\neg b = c$$
$$[\exists x[Fxc \wedge Fxb] \rightarrow c = b]$$

$\neg\exists x[Fxc \wedge Fxb]$	$c = b$
	$\neg b = b$ (by I)

At the bottom right, '$c = b$' is used to replace 'c' by 'b' in '$\neg b = c$', reaching the inconsistent formula '$\neg b = b$'.

Exercise 40D. Use tableaux to show that any argument which can be symbolized in one of the following ways must be valid.

1. Pb. *Therefore* $\forall x[x = b \rightarrow Px]$.
2. Pb. $\forall x \forall y[[Px \wedge Py] \rightarrow x = y]$.
   Therefore $\forall x[x = b \leftrightarrow Px]$.
3. $\forall x[x = b \leftrightarrow Px]$.
   Therefore $[Pb \wedge \forall x \forall y[[Px \wedge Py] \rightarrow x = y]]$.

Exercise 40E. Show that using Rule VII, there are closed tableaux for the following arguments:

1. $\forall xPx$. $\forall xQx$. *Therefore* $\exists x[Px \wedge Qx]$.
2. $[\forall xPx \vee \exists yQy]$. *Therefore* $\exists y[\forall xPx \vee Qy]$.

(There are no such tableaux without using Rule VII.)

The last few exercises of this section are a varied bunch; some are quite hard. You might be well advised to check your symbolizations before proceeding with the tableaux.

Exercise 40F. Prove the validity of each of the following arguments by translating them into symbols with the aid of the interpretation given, and then using tableaux.

Cx : x is a chimpanzee.
Sxy : x can solve y.
Px : x is a problem.
Txy: x is trying harder than y.
Bx : x will get a banana.
Mx : x is male.
b : Sultan
c : Chica

1. All the male chimpanzees can solve every problem. There's at least one problem. Any chimpanzee who can solve a problem will get a banana. Sultan is a male chimpanzee. *Therefore* Sultan will get a banana.
2. Sultan and Chica can solve exactly the same problems. If Sultan can solve any of the problems, then he will get a banana. Sultan will not get a banana. *Therefore* Chica can't solve any of the problems.
3. Not all the chimpanzees are trying equally hard. No chimpanzee is trying harder than himself. *Therefore* there are at least two chimpanzees.
4. Sultan is not Chica. Sultan won't get a banana unless he can solve all the problems. If the chimpanzee Chica is trying harder than Sultan, then Chica can solve a problem which Sultan can't. All the chimpanzees other than Sultan are trying harder than Sultan. *Therefore* Sultan won't get a banana.
5. Among all the chimpanzees, only Sultan is male. The chimpanzees who will get a banana are the males. *Therefore* Sultan is the chimpanzee who will get a banana.

---

41. Formalization Again†

From this point on, formalization of predicate logic can proceed much as in sections 22–25.

One can define a formal language L_2 whose grammatical sentences are those strings that could be got by analysing and symbolizing English predicates. The difference between free and bound occurrences of variables can also be described quite formally, and then we can pick out as *closed formulae* those grammatical sentences of L_2 that have no free occurrences of individual variables. Closed formulae are the formal counterpart of declarative sentences.

After formalizing the language, one tries to formalize situations and truth. It is possible to define L_2-*structures* by analogy with structures in section 23, so that a situation and an interpretation together determine an L_2-structure, and the L_2-structure in turn contains everything necessary for determining the truth-value of any closed formula in the situation and under the interpretation. Each L_2-structure is built around a set of individuals called its *domain*; the domain acts as domain of quantification for all quantifiers.

There are three main ways in which these structures are harder to handle than the structures in propositional logic.

First, one can have two completely different L_2-structures in which exactly the same closed formulae are true. This could never happen with the structures of section 23.

Second, L_2-structures can have infinitely many individuals in their domains.

Third, although L_2-structures do determine the truth-values of closed formulae, they do it in a way that tends to rule out direct calculation. In fact one can easily describe an L_2-structure which resists all systematic methods for calculating truth-values in it; if there were a method for computing truth-values in this structure, it would solve in one blow some arithmetical problems which have held out against three hundred years' battering by the world's best mathematicians.

Nevertheless there is a useful mathematical theory of L_2-structures; it goes by the name of *model theory*. One of the first steps in model theory is to copy definition (23.9) along the following lines. If X is a set of closed formulae and ψ is a closed formula, we write

† This section presupposes the mathematical sections 22–25.

$$X \models \psi \qquad \textbf{41.1}$$

to mean that there is no structure in which ψ and all the formulae of X are defined, and in which all the formulae of X are true but ψ is false. Just as before, expressions of the form (41.1) are known as *semantic sequents.*

Correct semantic sequents are a formal counterpart of valid arguments. We discussed the sense in which this is true at the end of section 23. Moving from propositional logic to first-order predicate logic allows us to capture very many more valid arguments than before. (But there is still no guarantee that every valid argument can be turned into a correct semantic sequent.)

For example, by the $\forall x\phi$ Rule of section 36, the following argument is valid:

$$\forall x\, x = x. \textit{ Therefore } a = a. \qquad \textbf{41.2}$$

Likewise we can use the $\exists x\phi$ Rule of section 36 to show that the argument

$$a = a. \textit{ Therefore } \exists x\, x = x. \qquad \textbf{41.3}$$

is valid too. So we should expect that the following two semantic sequents are correct:

$$\forall x\, x = x \models a = a. \qquad \textbf{41.4}$$

$$a = a \models \exists x\, x = x. \qquad \textbf{41.5}$$

And indeed it turns out that (41.4) and (41.5) are correct sequents.

On the other hand, the argument

$$\forall x\, x = x. \textit{ Therefore } \exists x\, x = x \qquad \textbf{41.6}$$

is not valid. Take the domain of quantification to be empty; for a possible example, let the situation be the real world and let the domain of quantification consist of all yetis. Then the closed formula '$\forall x\, x = x$' is true (remembering that we are using the weak reading from section 7), whereas '$\exists x\, x = x$' is false.

But now suppose the invalid argument (41.6) went over into an incorrect semantic sequent. Then comparing with (41.4) and (41.5), we can see that the transitivity of $\models$, Theorem IV in section 24, would no longer be true with our new definition of '$\models$'. This would be a real nuisance in the mathematical theory. Fortunately there is a way of preventing this nuisance: *we don't allow any L_2-structures with empty* *domains.* If we make this stipulation, sensible first-order predicate logic versions of all the Theorems of section 24 hold (except for XI, which is about truth-functors).

For technical convenience, many logicians make an adjustment in the rules for predicate tableaux so as to match this stipulation about structures. What is needed is a rule to guarantee that 'there is at least one thing'. Rule VII in section 40 has precisely this effect. There was also a quite different historical reason for adding Rule VII: the first proof calculi to be invented all allowed one to prove '$\exists x\ x = x$', so that Rule VII brings the tableau calculus into line with its predecessors.

How does this change affect the use of tableaux to prove the validity of arguments? The only difference will be that tableaux will pronounce that some arguments are valid when in fact they aren't. The arguments in question are those that become valid when we add

There is at least one thing (or person). ***41.7***

to the premises. If you like, you can think of this as an assumption about the kinds of argument that we are considering, just as the two Policies on Reference in section 28 were assumptions about the kind of sentence that we allow in the arguments that we are studying.

After these adjustments, one can define syntactic sequents for first-order predicate logic in full analogy to section 25, and one can prove the *Completeness Theorem*: a syntactic sequent is correct if and only if the corresponding semantic sequent is correct. Unfortunately the proof is too difficult even to sketch here.

Horizons of Logic

Having conquered **not**, **and**, **or**, **if**, **all**, **some**, **more** and **equals**, what next?

42. Likelihood†

> It'll probably be a boy. ***42.1***
> It'll probably be a girl.

These two sentences can't both be true at once – if it's probably a boy, then it's not likely to be a girl. So we have an inconsistency. Why?

One's first instinct is to blame the inconsistency of (42.1) on the fact that boys can't be girls. But this can only be part of the reason; for consider the sentences

> It'll probably be a boy. ***42.2***
> It'll be a girl.

The sentences of (42.2) are not inconsistent, because they can both be true at once. Improbable things do happen. It can happen that everything points to the child being a boy, when in fact it's going to be a girl; in such a situation (42.2) would all be true. (Of course we could never know at the time that we were in such a situation: if we knew the child was going to be a girl, it wouldn't any longer be probable that the child would be a boy.)

It appears that the word **probably** must share the blame for the inconsistency in (42.1). Shall we try to create a logic of **probably**?

† Parts of this section presuppose the mathematical section 23.

Perhaps the best starting-place is the 2-place sentence-functor

It's at least as likely that ϕ as that ψ. **42.3**

In symbols, we shall write (42.3) as '$[\phi \geqslant \psi]$'. Several English turns of phrase can be rendered fairly accurately with the help of (42.3). Here are some examples:

It's **more likely** that ϕ than that ψ. **42.4**
$\neg [\psi \geqslant \phi]$

More **likely** than not, ϕ. **42.5**
It's **likely** that ϕ.
Probably ϕ.
$\neg [\neg \phi \geqslant \phi]$

Certainly ϕ. **42.6**
$[\phi \geqslant [\phi \vee \neg \phi]]$

There's **no chance** that ϕ. **42.7**
$[[\phi \wedge \neg \phi] \geqslant \phi]$

It's **fifty-fifty** whether or not ϕ. **42.8**
$[[\phi \geqslant \neg \phi] \wedge [\neg \phi \geqslant \phi]]$

We want to determine what sets of sentences using the sentence-functor (42.3) are inconsistent. Unfortunately tableaux are no use to us here, since (42.3) is not a truth-functor. But the basic idea behind tableaux, namely to find conditions under which the sentences are true, may still serve.

Imagine the shuffled pack of cards in front of us. There are fifty-two cards arranged in four suits. Since the cards have been shuffled, they may be in any order. We may therefore be in any one of n different situations, where n is the number of ways a pack of cards can be ordered. Let us call these n situations the *allowed situations*. Now it is surely true that

The top card is more likely to be a diamond than a nine, **42.9**

Why is (42.9) true? Because the top card is a diamond in a quarter of the allowed situations, but it is a nine in only a thirteenth of these situations.

More generally, if the topic is the position of cards in the pack, then

It's at least as likely that ϕ as that ψ. **42.10** is true precisely if

> There are at least as many allowed situations in which it's true that ϕ as there are allowed situations in which it's true that ψ. ***42.11***

If (42.10) can be interpreted by (42.11) in this example, why not in others too? If so, then a formal calculus is easy to construct, assuming some ideas from the mathematical section 23 above.

We begin by constructing charts such as:

42.12

P	Q	R	measure
T	T	T	0.2
T	T	F	0.1
T	F	T	0.2
T	F	F	0
F	T	T	0.1
F	T	F	0
F	F	T	0.1
F	F	F	0.3

The chart (42.12) is to be interpreted as saying that among the allowed situations, the fraction in which 'P', 'Q' and 'R' are true is 0.2, the fraction in which 'P' and 'Q' are true while 'R' is false is 0.1, and so on. (The numbers on the right-hand side must add up to a total of 1.) A chart such as (42.12) is said to give a *probability measure.*

Using the chart (42.12), we can calculate the truth-value of any statement built up out of 'P', 'Q', 'R', '$\geqslant$' and truth-functors, in each of the structures listed in (42.12). For example we calculate the truth-value of '$[P \to [Q \geqslant R]]$' in each structure as follows.

42.13

P	Q	R	$[P \to [Q \geqslant R]]$
T	T	T	T F T F T
T	T	F	T F T F F
T	F	T	T F F F T
T	F	F	T F F F F
F	T	T	F T T F T
F	T	F	F T T F F
F	F	T	F T F F T
F	F	F	F T F F F
			(2) (1)

Column (1) is calculated thus: if we add up the total fraction of allowed situations in which 'Q' is true according to the probability measure (42.12), the answer comes to

$$0{\cdot}2 + 0{\cdot}1 + 0{\cdot}1 + 0 = 0{\cdot}4. \quad \textbf{42.14}$$

Likewise the fraction in which 'R' is true comes to

$$0{\cdot}2 + 0{\cdot}2 + 0{\cdot}1 + 0{\cdot}1 = 0{\cdot}6. \quad \textbf{42.15}$$

Hence 'Q' is true in fewer allowed situations than 'R', and so '$[Q \geqslant R]$' is false, according to (42.11). Column (2) is then calculated by the truth-table for '$\rightarrow$'.

A second example:

42.16

P	Q	R	$[Q \geqslant [P \wedge R]]$
T	T	T	T T T T T
T	T	F	T T T F F
T	F	T	F T T T T
T	F	F	F T T F F
F	T	T	T T F F T
F	T	F	T T F F F
F	F	T	F T F F T
F	F	F	F T F F F

To calculate the column below '$\geqslant$', we first recall from (42.14) that 'Q' is true in 0.4 of the allowed situations. Then we calculate the fraction in which '$[P \wedge R]$' is true, again from (42.12):

$$0{\cdot}2 + 0{\cdot}2 = 0{\cdot}4. \quad \textbf{42.17}$$

Since 0.4 is at least as great as 0.4, the whole is true.

For convenience we can refer to (42.13) and (42.16) as *truth-tables*. But there are two significant differences between them and the truth-tables of section 23. First, the truth-value of a formula may now depend on the choice of some probability measure which has to be supplied separately. Second, the truth-value of a formula in one structure may now depend on the truth-values of certain formulae in other structures; we have caught a situation-shifter red-handed.

Exercise 42. Using the probability measure (42.12), give truth-tables for each of the following formulae:

1. $[R \geqslant \neg R]$
2. $\neg [[Q \wedge P] \geqslant [Q \wedge \neg P]]$
3. $[R \geqslant [P \geqslant Q]]$

We call a formula *valid* if it's true in every structure and for every probability measure. If we are on the right lines so far, the valid formulae will be the formal counterparts of the necessary truths involving the sentence-functor (42.3). It is not hard to establish which formulae are valid, though it needs more mathematics than we have space for.† One can show for example that all the following formulae are valid:

$$[P \geqslant P] \qquad \textbf{\textit{42.18}}$$
$$[[[P \geqslant Q] \wedge [Q \geqslant R]] \rightarrow [P \geqslant R]]$$
$$[[P \geqslant Q] \vee [Q \geqslant P]]$$
$$[[P \geqslant Q] \rightarrow [\neg Q \geqslant \neg P]]$$

The second formula of (42.18) assures us that when 'it' has been specified, the following sentence must necessarily be true:

> If it's at least as likely to be a cold as flu, and at least as likely to be flu as measles, then it's at least as likely to be a cold as measles. ***42.19***

The other formulae can be similarly interpreted. Pursuing the analogy with section 23, one can also devise correct semantic sequents and interpret them into valid arguments.

This logic of likelihood is elegant and convincing; it meshes well with the mathematical theory of probability. Nevertheless our account of it has one major flaw which we must correct.

The flaw is a matter of interpretation. At the beginning of our discussion we imagined a shuffled pack of cards, and we noted that as far as the order of the pack was concerned, we could be in any one of several possible situations. Without saying so explicitly, we assumed that *every one of these situations was equally likely to be the actual situation*. If some

† See Peter Gärdenfors, 'Qualitative Probability as an Intensional Logic', *Journal of Philosophical Logic*, 4 (1975), pp. 171–85.

orderings of the pack had been more likely than others – if, say, there was a suspicion that a clever shuffler had brought all the nines to the top – then (42.9) need no longer be true. In a similar way, the restriction to 'allowed' situations merely excluded those situations which we thought were too unlikely to take seriously. For example, if there was a chance that someone had quietly removed the diamonds from the pack, we would have to take this into account when we assessed the truth of (42.9).

For reasons such as these, we are forced to reinterpret probability measures as showing, not what proportion of the allowed situations assign the indicated truth-values, but *how likely it is* that the actual situation is one which assigns them. Once this reinterpretation is made, our calculations can no longer be justified by an appeal to (42.11). Instead we have to base them on our intuitive understanding of probability.

In other words, our calculus only tells us how to deduce likelihoods from other likelihoods. It is not clear how far likelihoods can be deduced from anything else. Certainly we all do estimate likelihoods all the time – what chance I'll reach the shops before closing time? Might James take offence? Will another drink make me feel sick? Is the ladder safe? Apparently we do it on the basis of the facts we know. But nobody has yet provided a complete and convincing account of how to deduce a likelihood from brute facts alone. Maybe it can't be done, and estimating likelihoods is fundamentally different from deducing them. Maybe it can, but only by arguments which are too long to set down on paper.

Here is an example of the difficulties. It used to be thought that if every S so far discovered is a P, then the discovery of one more S which was a P would make it more likely that every S is a P. But in fact this is not so. For instance, take the question whether recognizably human animals (hominids) existed four million years ago. Every hominid discovered so far is less than 3.8 million years old. On the earlier view, the discovery of another hominid skull less than four million years old should make it *less likely* that hominids go back four million years. But in fact if a new hominid skull was found and dated to 3.9 million years ago, this would make it *more likely* that our hominid ancestors go back still further.

Likewise, if both the known sufferers from a rare bone disease are called John, then the discovery of a third John with the disease would make it more likely that there are people not called John who have the disease.

43. Intension

The attempt to form a logic of likelihood presented us with some difficulties, but there was a useful practical hint among them. Although the truth-value of a sentence of the form '$[\phi \geqslant \psi]$' in a given situation couldn't be calculated from the truth-values of ϕ and ψ in that one situation, it could be worked out once we knew the truth-values of ϕ and ψ in each and every situation. (We did also have to be given a probability measure to indicate how likely the various situations were; but this was independent of the sentence under discussion.)

By the *intension* of a declarative sentence, we mean a correlation between situations and truth-values, which correlates to each situation the truth-value which the given sentence has in that situation. In particular, two sentences have the same intension precisely if they have the same truth-value as each other in every situation. All necessary truths have the truth-value Truth in all situations, so that all necessary truths have the same intension. (In essentially this sense, the term comes from Rudolf Carnap, 1947; any resemblance to intentionality with a 't' is an unfortunate coincidence.)

We can say, then, that the truth-value of '$[\phi \geqslant \psi]$' in a situation is determined by the intensions of ϕ and ψ. It may be possible to represent the intensions of ϕ and ψ by columns of 'T's and 'F's in a truth-table; the truth-value of '$[\phi \geqslant \psi]$' then becomes a matter of straightforward calculation, as we saw in section 42.

Analogy suggests a similar definition for designators. The *intension* of a designator D is a correlation between situations and things; to each situation in which D has a primary reference, the intension correlates this primary reference, while it correlates nothing to the remaining situations. To describe the intension of a designator, we can list, for each situation, whether the designator has a primary reference in the situation, and if so, what. For example, the intension of the designator **the monarch of England** would need entries such as the following, with the dates made more precise:

43.1

situation	primary reference
1603–25	James I of England
1625–49	Charles I
1649–60	none – no monarch
1660–85	Charles II
1685–88	James II
1688–95	none – two monarchs

One can define *intension* in the same spirit for other kinds of expression too. For example, the *intension* of a predicate correlates to each situation (which is here supposed to include a domain) the relation which is expressed in that domain and situation by that predicate. So to describe the intension of the predicate '*x* is red' we need a chart that shows, for every situation, all the things that are red in that situation.

Because the truth-value of a sentence about likelihoods was determined by the intensions of constituent sentences, we were able to make a logic of likelihood. We should expect the same to hold quite generally: *if the truth-value of a sentence in a situation is determined in a describable way by the intensions of its constituents, then a system of logic is there to be discovered.*

Shall we look around and see what we can find?

(i) *Likelihood*

We have already discussed this.

(ii) *Necessity and possibility*

We can consider together a number of kinds of sentence, which are true in a situation precisely if some shorter sentence is true in *every one* of certain allowed situations. (These allowed situations are usually said to be *accessible.*) Examples are:

It's a **necessary truth** that one plus one is two. ***43.2***

(43.2) is true precisely if 'One plus one is two' is true in every possible situation.

Bob **couldn't** stop himself laughing. ***43.3***

(43.3) is true precisely if 'Bob is laughing' is true in every situation which Bob was capable of bringing about.

The needle **must be** somewhere in this room. ***43.4***

(43.4) is true precisely if 'The needle is in this room' is true in every situation which is compatible with what I know about the needle and the room.

The screw is **too** big to fit through the hole. ***43.5***

(43.5) is true precisely if in every possible situation in which the hole has its present size, it's true that no screw as big as the present one is going through the hole.

Each of these sentences has a negative back half:

It's **possible** that one plus one is not two. ***43.6***
Bob was **able** to stop himself laughing. ***43.7***
The needle **needn't** be anywhere in this room. ***43.8***
The screw is small **enough** to fit through the hole. ***43.9***

The logic of necessity is called *modal logic*, and it has been widely studied.† You can discover some of it for yourself now if you try to rewrite section 42, replacing the sentence-functor (42.3) by the sentence-functor

It's a necessary truth that ϕ. ***43.10***

(The sentence-functor (43.10) is commonly symbolized '$\Box\phi$'.)

(iii) *Tensed statements*

Some sentences about the past or future are true in the present if and only if certain other sentences were or will be true at certain other times. For example:

I have **always** been an idealist. ***43.11***

is true now if and only if the sentence 'I am an idealist' was true in every past situation in which I was anything.

I **shall** be in Oswestry tomorrow. ***43.12***

is true now if there is a time tomorrow at which the sentence 'I am in Oswestry' will be true.

I **often** get cramp. ***43.13***

is true now if at several times in the recent past the sentence 'I've got cramp' was true.

Henry **was** a prude **until** he married. ***43.14***

is true now if and only if there is a past situation in which 'Henry is just married' was true, such that in every earlier situation involving Henry, 'Henry is a prude' was true.

These four examples suggest that all tenses should be thought of as situation-shifters which change the relevant situation from the present to

† G. E. Hughes and M. J. Cresswell, *A New Introduction to Modal Logic*, Routledge 1996.

past or future ones. On this basis one can build up a logic of tenses, again very much as the logic of likelihood in section 42†. But here is an example which shows that unexpected subtleties may be involved.

Consider the sentence

I have read Gibbon's *Decline and Fall.* ***43.15***

There is no present-tense English sentence ϕ such that (43.15) is true now if and only if ϕ was true at some time in the past. The sentence

I am reading Gibbon's *Decline and Fall.* ***43.16***

will not do, because it carries no implication that I shall finish the book, while (43.15) definitely implies that I have finished it. If we insist on regarding (43.15) as the past-tense version of some present-tense sentence, we shall have to invent a new kind of present-tense sentence in order to do it.

44. Semantics

We saw that propositional logic rests on the meanings of English words such as **and**, **or** and **not**. Later we learned that predicate logic springs from the meanings of English words such as **every**, **some** and **equals**. In the last section we discovered that there are logics based on the meanings of the English words **possible**, **enough** and **until**.

That brings two questions to mind. First, does every English word have its own kind of logic? And second, is logic really a branch of linguistics?

The answer to the second question is certainly No. The formula

$[P \to [Q \to P]]$ ***44.1***

is a tautology and would have been a tautology even if the human race and its languages had never existed. The fact that it is a tautology is a fact of logic and not a fact about any natural language. Of course to *apply* (44.1) *in English arguments* we need to know that certain sentences of English can be paraphrased into the form (44.1) with the help of suitable interpretations; and at that point we need to know something about English. But if this book had been written in Hindi or Xhosa or Quechua,

† R. Goldblatt, *Logics of Time and Computation*, CSLI, Stanford University, 1987.

it could have covered the same facts of logic without containing any English at all.

In fact one of the main achievements of nineteenth and twentieth century logic was to isolate a robust mathematical core of logic that is independent of any language. In earlier centuries, logicians tried to state their laws for sentences of a particular natural language. The result was that all their laws had dozens of exceptions, and this helped to give logic a bad name.

The first question takes longer to answer. We shall see that meanings do give rise to logics in a fairly systematic way, but there is no guarantee that many of these logics will be interesting. The study of meanings is called *semantics*. The rest of this section will explore some connections between semantics and logic.

Consider two principles about meaning:

> *Substitution Principle.* Suppose X and Y are two words that have the same phrase-class (see section 12) and mean the same, S is a sentence containing X, and S is true in a certain situation. If we replace X by Y at one or more places in S, then the resulting sentence is still true in the given situation.

For example **courgette** means the same as **zucchini**. So if it's true that my aunt in Budleigh Salterton was poisoned by a courgette last Thursday evening, it's also true that my aunt in Budleigh Salterton was poisoned by a zucchini last Thursday evening.

> *Non-substitution Principle.* Suppose X and Y are two words that have the same phrase-class but don't mean the same. Then there is a sentence S containing X, and a situation, such that S is true in the situation but we can turn S into a false sentence (in that situation) by replacing X at one or more places by Y.

This principle says that the method of Exercise 8, for showing up differences in meaning, always works.

Whether or not these two principles are a hundred per cent true, they suggest an interesting way of doing semantics. We look for rules that determine the conditions under which any given sentence is true. If we have these rules, then we can look at a word and see how the rules use it when it occurs in a sentence. The two principles together suggest that a pair of words (of the same phrase-class – we need not keep repeating this) have the same meaning if and only if the rules treat them alike.

This should sound familiar. In propositional logic we examined when a given sentence is true in a given situation. The method was to start by finding the truth-value of each short constituent sentence in the situation, and then use the truth-table rules to work up through longer and longer constituent sentences until we found the truth-value of the whole sentence. Imagine this method rewritten so that it applies to all situations at once. At the start we would assign to each short sentence its intension, i.e. a specification of what situations it is true in. Then we would use truth-table rules to work out the intensions of all the constituent sentences, until we reached the intension of the original sentence. The intension of that sentence tells us the situations in which the sentence is true.

A semantics that goes this way, by assigning some object to each word in a sentence and then assigning objects in turn to longer and longer constituent phrases until eventually an object is assigned to the whole sentence, is said to be a *denotational semantics*. The object assigned to a word or a phrase is called the *denotation* of the word or phrase. (This is a technical term. Don't try to connect it with other uses of the word **denote**.) In the semantics that we gave for sentences built up using **and**, **not**, **until** and so forth, we used intensions as denotations.

Suppose we have a denotational semantics for the whole of English. Then the two principles above should lead us to expect that a pair of words have the same meaning if and only if they have the same denotation. So denotations are a tool for studying meanings. But they need not be the same thing as meanings.

In fact the intension of a sentence is certainly not the same thing as the meaning of the sentence, in any ordinary use of the word **meaning**. For example the following two sentences are necessary truths, and hence they have the same intension (because they are both true in all possible situations):

$1 + 1 = 2.$ ***44.2***

$7,536,429 + 8,897,698 = 16,434,127.$

Nevertheless the sentences obviously don't mean the same thing, since one of the following sentences is true and the other is false:

It's obvious that $1 + 1 = 2$. ***44.3***

It's obvious that $7,536,429 + 8,897,698 = 16,434,127$.

This example shows more: it shows that intension won't work as a denotation for sentences if we try to construct a denotational semantics for sentences that contain the word **obvious**. (There is not yet any generally accepted denotational semantics covering the word **obvious**. The verbs **believes** and **knows** give similar problems.)

So the word **obvious** destroys any semantics that uses intensions as denotations. This illustrates a common phenomenon in denotational semantics, as in the next example.

The sentence

Alberto's handwriting is legible. ***44.4***

is true in a situation if and only if, in that situation, Alberto's handwriting is in the class of things that are legible. So in determining whether (44.4) is true, the information that we need from the word **legible** is the class of things that are legible. This suggests that we take as the denotation of **legible** the assignment to each possible situation of the class of things that are legible in that situation. In fact this suggestion works well and gives a good semantics for interpreting predicate logic. (The L_2-structures of section 41 are built around it.)

At least, it works well until we remember section 32 and add the word **more** to the language. Then we shall have sentences such as

Alberto's handwriting is more legible than Bernardo's. ***44.5***

It's possible that Alberto's handwriting is beautifully clear but Bernardo's is only just legible. In such a situation, (44.5) is true although both handwritings are legible. So the truth of (44.5) depends on more than distinguishing between legible and illegible; it depends on there being a *scale of legibility*. Then (44.5) is true if and only if Alberto's handwriting comes higher on the scale than Bernardo's does. The denotation of **legible** must tell us what this scale is in each situation. That's more than our suggestion in the previous paragraph allowed us.

Note carefully what has happened here. By extending our semantics to sentences containing the word **more**, we add a complication to the semantics of other words such as **legible**. In fact we complicate the semantics of all adjectives.

These examples strongly suggest that there is no hope for devising a denotational semantics for English piecemeal. Add one word and you may need to rejig the whole of the rest of the system. It's a task for Sisyphus.

Nevertheless one can find good denotational semantics for quite substantial fragments of English. How does this connect with logic? Very directly: if we have a semantics that tells us in which circumstances certain sentences are true, then it will tell us (at least in principle) which sets of those sentences are never all true in the same situation; in other words it will tell us which sets of sentences are inconsistent. But this was exactly what we needed in order to do logic.

Does this tell us that every denotational semantics gives rise to a useful logic? Not necessarily – the rules of the semantics might be too unwieldy to apply in any useful way. There is another problem too. The Substitution Theorem in section 24 showed that if we can prove the validity of an argument by propositional logic, then (to quote section 24)

> we can . . . replace each occurrence of some sentence in the argument by another sentence (the same throughout), without destroying the validity.

A similar remark holds for predicate logic too. These two logics don't just prove the validity of particular arguments; they prove the validity of *all arguments of a given form.*

So a more extended semantics may do very little for logic if it doesn't yield new valid *forms of argument.* For example, does the word **obvious** give us new valid forms of argument? Well, it seems to give us one:

> It's obvious that *P. Therefore P.* **44.6**

People have suggested others, but they are less convincing. For example:

> It's obvious that *P. Therefore* it's obvious that it's obvious that *P.* **44.7**

There is a story about the mathematician G. H. Hardy, that when he said in a lecture that some equation was 'obvious', someone in the class protested that it wasn't. So Hardy walked up and down the corridor for five minutes while he thought about it. Then he said to the class: 'Yes it is obvious. I proceed.'

From (44.7) one might draw the moral that words like **obvious** are too subtle or vague to allow any significant number of general laws. This is the wrong moral. Before the development of propositional logic, it would have been easy to point to all kinds of subtleties in the use of the words

and, **or** and **not** (including those mentioned in sections 7 and 17) as reasons why there couldn't be a significant logic of sentences.

Our discussion leaves it entirely open whether future developments in semantics are likely to produce interesting new logical calculi. It would be foolhardy to make any predictions. Two hundred years ago, nobody foresaw either propositional or predicate logic.

Answers to Exercises

Section 1

1A. If you think you succeeded, you probably overlooked the difference between believing a thing and imagining it.

1B.
1. Consistent, but highly improbable.
2. As it stands, inconsistent; you can't literally know a thing that isn't true. Maybe the speaker only meant that she felt certain she would never get pregnant.
3. As it stands, consistent; the people with nowhere to live might be people in Glasgow or Jersey City. Nevertheless there seems to be some serious confusion in the mind of the Mayor of Lincoln who made this ripe remark to the annual dinner and dance of Lincoln and District Association of Building Trades' Employers (as quoted on p. 70 of *This England*, New Statesman, 1960).
4. Inconsistent. Did Walter pay for his own holiday? If he did, then he was a club member who did pay for his own holiday; so he didn't pay for his holiday. If he didn't, then he was a club member who didn't pay for his own holiday; so he did pay for his holiday. We reach an absurdity either way. (This is a version of Bertrand Russell's paradoxical barber, who shaved the villagers who didn't shave themselves.)
5. This gāthā of the Ch'an Master Fu Ta Shih (quoted from p. 143 of Lu K'uan Yü, *Ch'an and Zen Teaching* (*Series One*), Rider & Co., 1960) was surely meant to be inconsistent, and I think it succeeds. Followers of Ch'an, or its Japanese version Zen, regard sayings like this as fingers pointing to a reality which is normally obscured by the daily round of emotions and calculations.

Section 3

3A.

1. Selection mistake.
2. Hopeless. Noam Chomsky thought up the two sentences 1 and 2 in his book *Syntactic Structures*, Mouton, The Hague, 1957, to illustrate a point about degrees of grammaticality. They have been frequently quoted since, for many purposes.
3. Perturbation of 'Please pass me some butter' or 'Please pass me the butter.'
4. Usually reckoned grammatical but informal.
5. This charming selection mistake is from e.e. cummings, *my father moved through dooms of love.*
6. From the same poem; hopeless, I'd say.
7. In many people's English this sentence (from a letter of the American writer Robert Frost, quoted by *Webster's Dictionary of English Usage* under 'double negative', is grammatical but informal.
8. Perturbation of 'His face was calm and relaxed, like the face of a sleeping child' or '. . . a child asleep'.
9. Perturbation of 'Father, do not prolong further our necessary defeat.' This is the opening couplet of a poem of W. H. Auden, presumably in mock imitation of the hymn-writers Tate and Brady, who in 1696 published verse translations of the Psalms. These two authors made grammatical perturbations their trademark. Specimen:

   Deliverance he affords to all
   Who on his succour trust.

10. Selection mistake, from *Burnt Norton*, T. S. Eliot.

3B.

1. Declarative sentence, as are all mathematical statements. Here and below, we shall overlook the fact that mathematical and scientific statements often contain symbols and technical terms which are not in normal use in English.
2. Not declarative. Requests are not declarative.
3. Not declarative, although it expresses very much the same as the declarative sentence 'You're standing where Cromwell once stood.'
4. Declarative.

5. This is controversial. Those who believe it is *not declarative* say that the sentence is not used to state a belief but to make a promise. (The same would apply to sentences beginning 'I swear . . .', 'I congratulate. . .', 'I forbid . . .', etc.) On this view, 'Is it true that I promise not to peep?' would presumably contain some kind of selection violation, or worse. Those who believe 5 is *declarative* would argue that 'He promises not to peep' is undoubtedly declarative, and 5 is hardly a different kind of sentence.
6. Controversial, rather like 5. Most people would regard it as declarative. But some philosophers have argued that if a person says it, she isn't really making a statement at all; rather she is giving vent to strong feelings of disapproval of blackmail. This view tends to diminish the relevance of logic to moral thinking.

Section 4

4A.

1. Cross-reference: '. . . *her* stern as *she* slid gracefully down the slipways'.
2. Lexical: '*pulled . . . through a difficult passage*' meant literally or metaphorically.
3. Both lexical and structural. (1) cases = instances, and he admitted two cases. (2) cases = containers, and he admitted stealing two cases.
4. Cross-reference: both ends of what?
5. Lexical: (1) bare bones = essentials, (2) bare bones = uncovered pieces of skeleton.
6. Structural: (1) were strongly urged, by the Liverpool justices, to take steps; (2) were strongly urged to put a stop to the evil, namely, Liverpool justices drinking methylated spirits.

   (The items in this exercise were taken from Denys Parsons, *Funny Amusing and Funny Amazing*, Pan, 1969, with one slight alteration.)

4B.

1. a. I shall wear no clothes that would distinguish me from my fellow citizens.
   b. To distinguish me from my fellow citizens, I shall wear no clothes.
2. a. He never relaxes except on Sundays.
   b. On Sundays he does nothing but relax.

3. a. He gave each guest either a glass of rum or a glass of gin and tonic.
   b. He gave each guest either a glass of rum and tonic or a glass of gin and tonic.
4. a. There is one person who has in his or her hands most of the nation's assets.
   b. Most of the nation's assets are in the hands of individuals. (Not perfect, but hard to improve in less than five lines.)
5. a. If you have a dog you must carry it.
   b. You must have a dog and carry it.

Section 5

5. he = the man; he = the man; him = Jacob; he = the man; he = Jacob?; him = the man?; he = the man; me = the man; he = Jacob; I = Jacob; thee = the man; thou = the man; me = Jacob; he = the man; him = Jacob; he = Jacob; he = the man; thou = Jacob; him = the man; me = Jacob; I = Jacob; thee = the man; he = the man; thou = Jacob; he = the man; him = Jacob.

Section 6

6. 1. scaling. 2. scaling. 3. not scaling. 4. not scaling. 5. scaling. 6. scaling. 7. not scaling. 8. not scaling. 9. scaling. 10. not scaling. (Some may be controversial.)

Section 8

8.1. The alderman had been asked how much it cost to buy the land for the new school. He answered in good faith by reporting what he himself had been told, namely that it cost half a million pounds. However, his informant had misread the records, and the true cost was three-quarters of a million. (First sentence of exercise false, second true.)

2. If your creditors catch up with you they will ruin you. Never mind; a friend of mine has a place in Corsica where you can go underground for a year or so, and I can fix it up for you. It's a shame that the Corsican food doesn't suit your digestion. (First sentence true, second false.)

3. I'm not normally at home at this hour, because it's my evening for Yoga. But they cancelled the class, so I stayed at home. The man watching the house had been told I was at the class; but he saw Tim

walking into the house just now, and he thought it was me coming home from the class. Tim and I are often mistaken for each other. (First sentence false, second true.)

4. Brutus mentioned to Servilius Casca that Tillius Cimber had sworn to pay a large sum of money to anybody who would kill Caesar. This inspired Casca – who had gambling debts – to kill Caesar by stabbing him through the heart. (First sentence false, second true.)
5. We anthropologists are learning the local Bushman language fast. We have just learned that 'crow' is their word for a daddy longlegs. Incidentally I showed some of them a photo of a carrion crow, and they said the bird must have been painted. (First sentence true, second false.)

Section 9

9A. All inconsistent except 2 and 6.

9B. 1. With 'can'; inconsistent.
2. With the phrase that follows; consistent.
3. With 'could'; inconsistent.
4. With the phrase that follows; consistent.
5. With 'did'; inconsistent.

Section 10

10.1

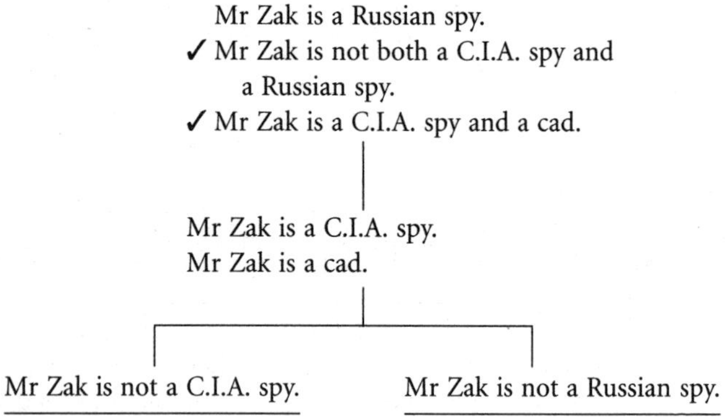

Mr Zak is a Russian spy.
✓ Mr Zak is not both a C.I.A. spy and a Russian spy.
✓ Mr Zak is a C.I.A. spy and a cad.

|

Mr Zak is a C.I.A. spy.
Mr Zak is a cad.

Mr Zak is not a C.I.A. spy.	Mr Zak is not a Russian spy.
___	___

A closed tableau; therefore *inconsistent.*

2.

✓ At least one of Auguste and Bruno lives in Bootle.
✓ At least one of Bruno and Chaim is an estate agent.

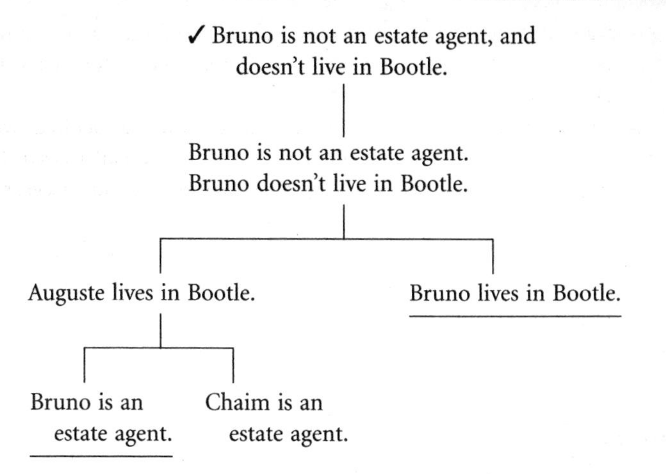
✓ Bruno is not an estate agent, and doesn't live in Bootle.
Bruno is not an estate agent.
Bruno doesn't live in Bootle.
Auguste lives in Bootle.
Bruno lives in Bootle.
Bruno is an estate agent.
Chaim is an estate agent.

There is an unclosed branch; it describes the possible situation that Auguste lives in Bootle, Bruno doesn't live in Bootle and is not an estate agent, and Chaim is an estate agent. In this situation the three sentences of the exercise are all true. Therefore they are *consistent.*

3.

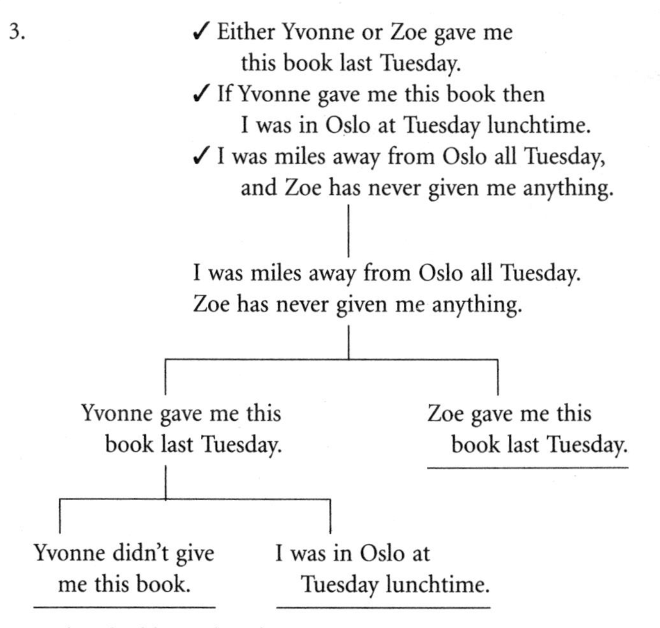
✓ Either Yvonne or Zoe gave me this book last Tuesday.
✓ If Yvonne gave me this book then I was in Oslo at Tuesday lunchtime.
✓ I was miles away from Oslo all Tuesday, and Zoe has never given me anything.
I was miles away from Oslo all Tuesday.
Zoe has never given me anything.
Yvonne gave me this book last Tuesday.
Zoe gave me this book last Tuesday.
Yvonne didn't give me this book.
I was in Oslo at Tuesday lunchtime.

A closed tableau; therefore *inconsistent.*

Section 11

11.1. Two hundred people are dying every day. *Therefore* help is needed urgently.

2. When Communists operate as a minority group within unions, settlements by the established officials must be denounced as sellouts. *Therefore* strikes are unlikely to wither away in any democratic country so long as Communists have strong minority influence.
3. The nests of the verdin are usually placed at or near the end of a low branch. *Therefore* the nests of the verdin are surprisingly conspicuous.
4. The effect of salicylates on renal uric acid clearance is greater than the effect which ACTH has on it. *Therefore* the effect of ACTH on gout is not due to the increased renal uric acid clearance alone.
5. If no contribution to the magnetic field comes from electric currents in the upper atmosphere, then we cannot account for the relation between the variations in the magnetic elements and the radiation received from the sun. *Therefore* some contribution to the magnetic field comes from electric currents in the upper atmosphere.

Section 12

12.

	health	a wife	wife	cameras	two cameras
He wanted to have *x*.	✓	✓	×	✓	✓
He wanted to have the *x*.	×	×	✓	✓	?
He wanted to have more *x*.	✓	×	×	✓	×

The question-mark in the last column reflects my doubt about whether 'two cameras' is a constituent of the grammatical sentence 'He wanted to have the two cameras.'

Section 13

13A. 1. both pigeons revealed surprising adaptability

2. S, NP, VP, Det, N, V, Adj.

3. 1 [both pigeons revealed surprising adaptability]
   2 [both pigeons] revealed surprising adaptability
   3 [both] pigeons revealed surprising adaptability
   4 both [pigeons] revealed surprising adaptability
   5 both pigeons [revealed surprising adaptability]
   6 both pigeons [revealed] surprising adaptability
   7 both pigeons revealed [surprising adaptability]
   8 both pigeons revealed [surprising] adaptability
   9 both pigeons revealed surprising [adaptability]

13B. Most linguists would give the phrase-marker

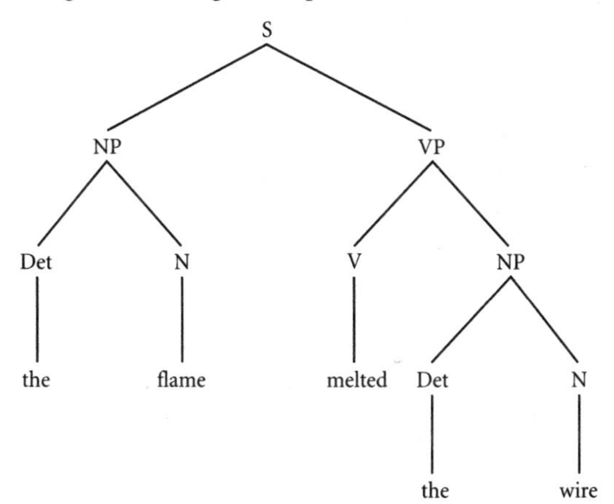

The constituents are then as follows:

1 [the flame melted the wire]
2 [the flame] melted the wire
3 [the] flame melted the wire
4 the [flame] melted the wire
5 the flame [melted the wire]
6 the flame [melted] the wire
7 the flame melted [the wire]
8 the flame melted [the] wire
9 the flame melted the [wire]

But the following phrase-marker is not definitely wrong:

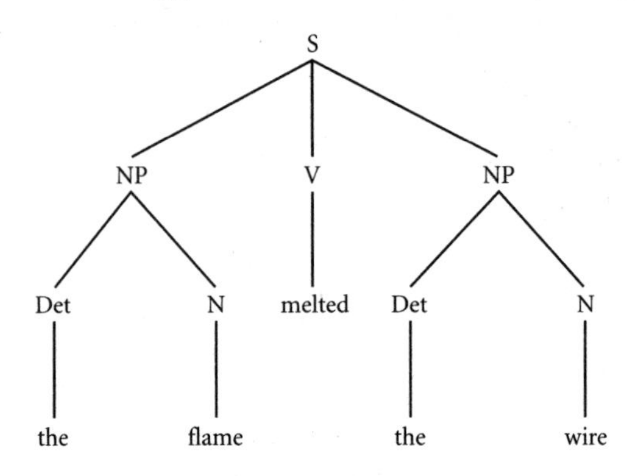

The constituents are then the same as before, except that number 5 disappears.

13C. Several details of the phrase-markers given below could well be different. But your answers should have the stated constituents.

1.

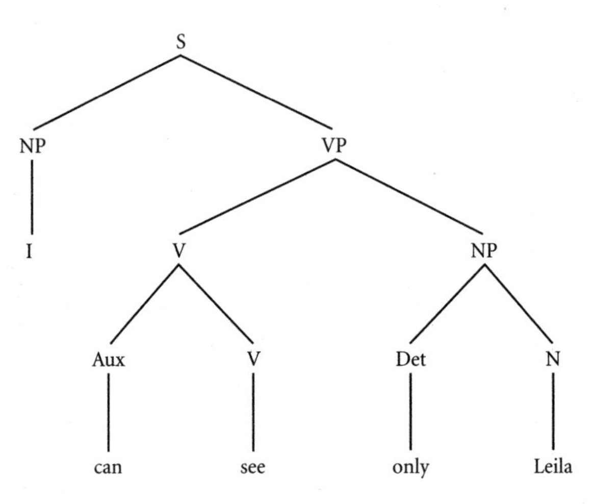

'only Leila' is thus an underlying constituent. (Aux = Auxiliary.)

2. As a first approximation,

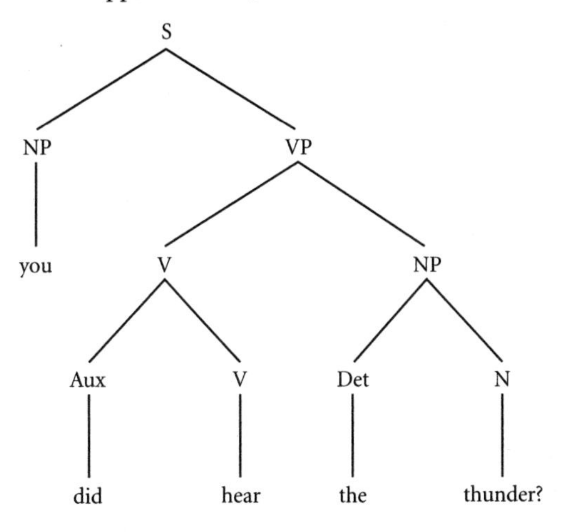

'did hear' is thus an underlying constituent. The question-mark is sometimes taken as a separate constituent too.

3.

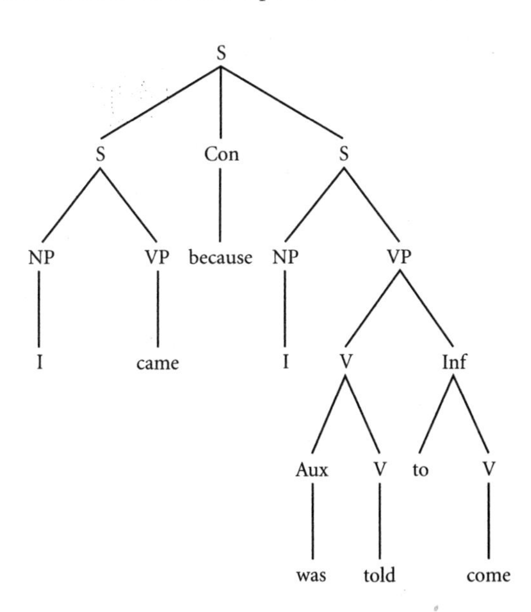

'to come' is thus an underlying constituent.

4.

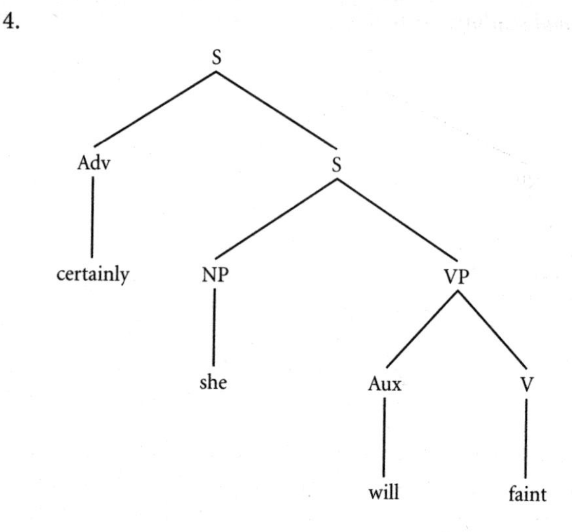

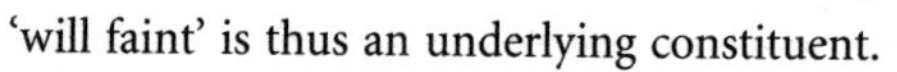

'will faint' is thus an underlying constituent.

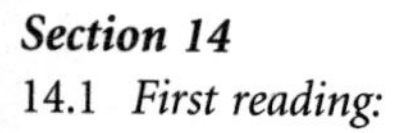

Section 14

14.1 *First reading:*

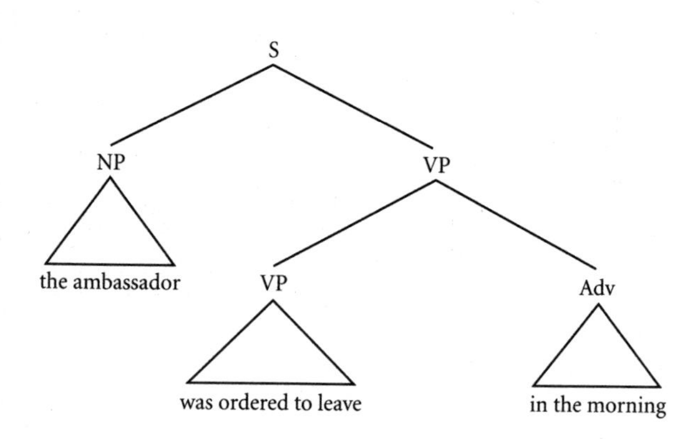

Second reading:

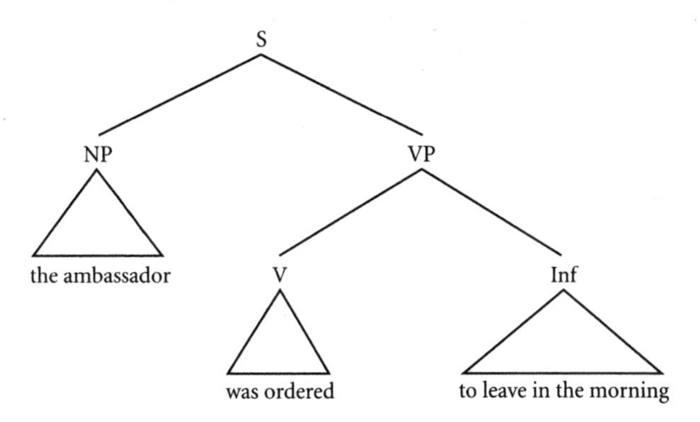

Alternatively you could attach 'in the morning' to the whole sentence 'the ambassador was ordered to leave' in the first reading. The central point is that in the second reading, the scope of 'in the morning' is 'to leave in the morning', while in the first reading the scope of this phrase includes 'was ordered'.

2. *First reading:*

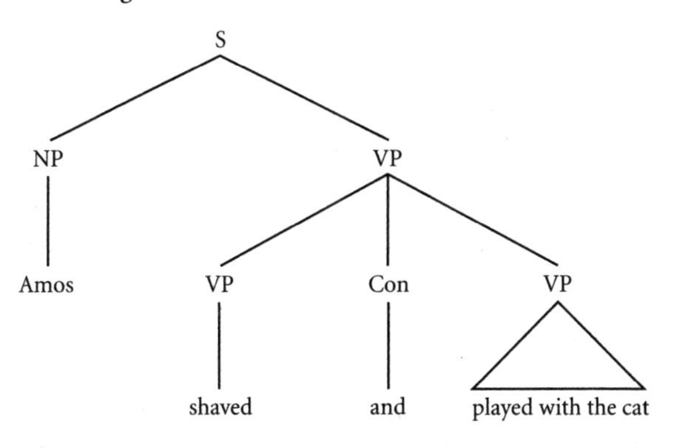

Second reading:

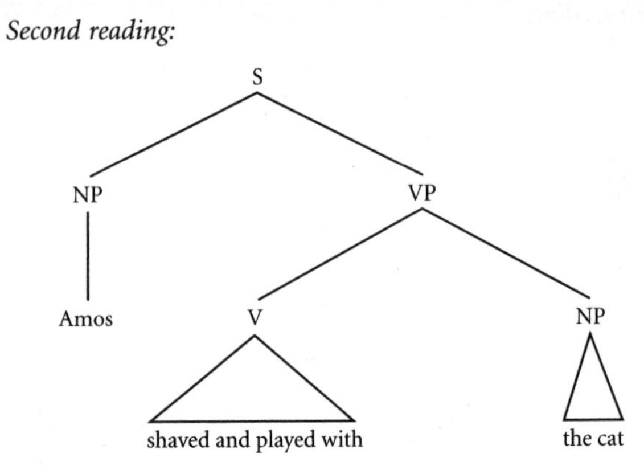

It seems likely that the second reading has an underlying phrase-marker

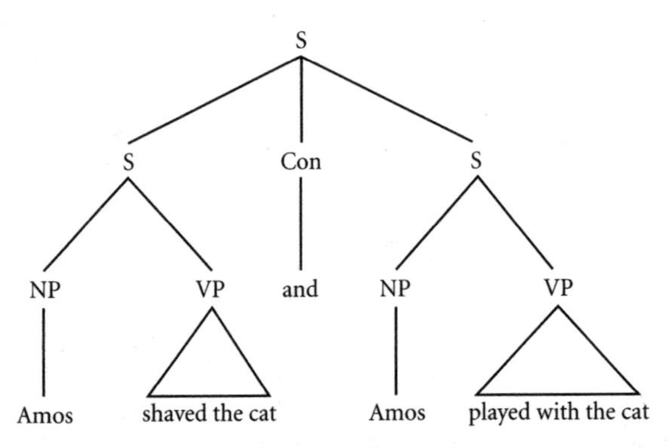

or something similar; note that 'shaved the cat' is a constituent in this underlying phrase-marker.

3. *First reading:*

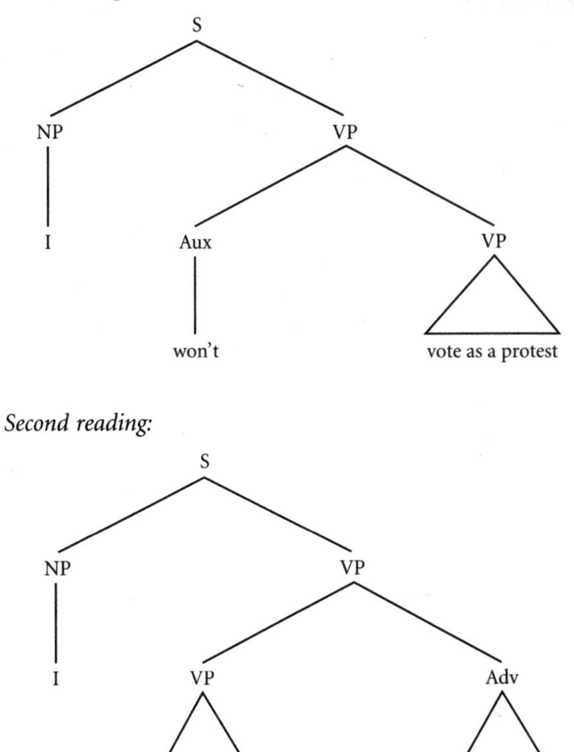

Second reading:

Note that the scope of 'won't' is 'won't vote as a protest' in the first reading, but only 'won't vote' in the second.

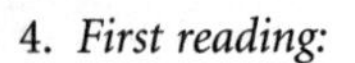

4. *First reading:*

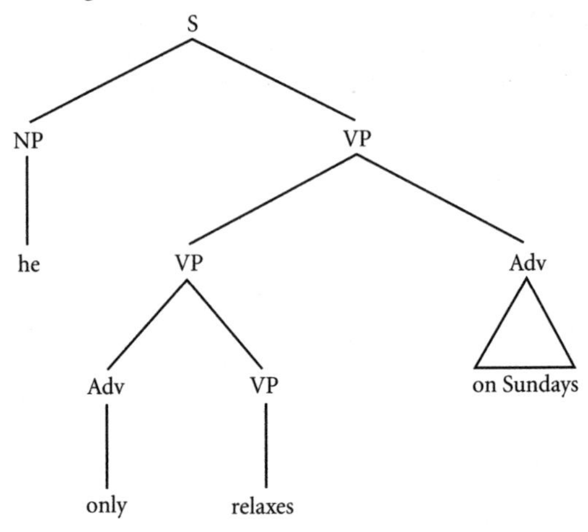

The *second reading* needs an underlying phrase-marker with 'only' shifted along so that its scope is 'only on Sundays':

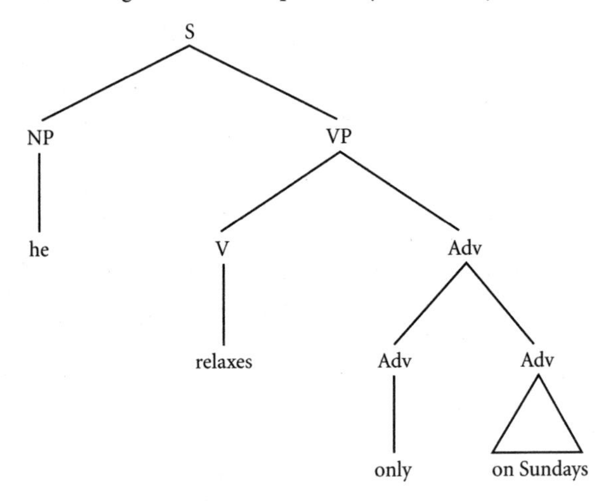

Section 15

15A. Initial symbol: S
Non-terminal symbols: S, VP, NP, N, Det, Adj, V.
Terminal symbols: that, the, white, man, cat, resembles.

15B. 1.

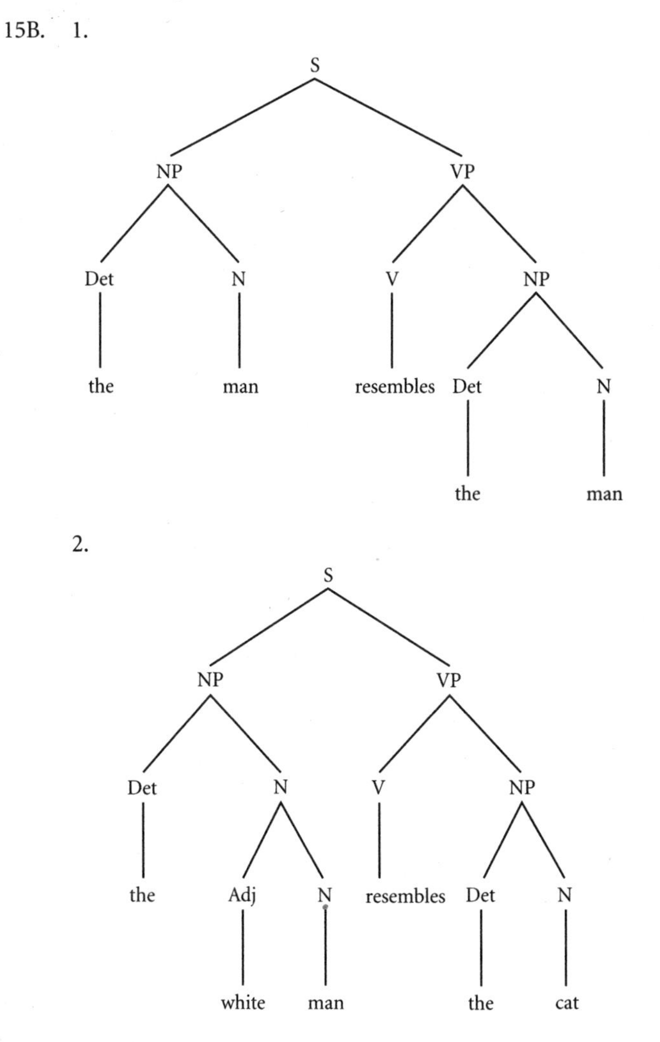

3.

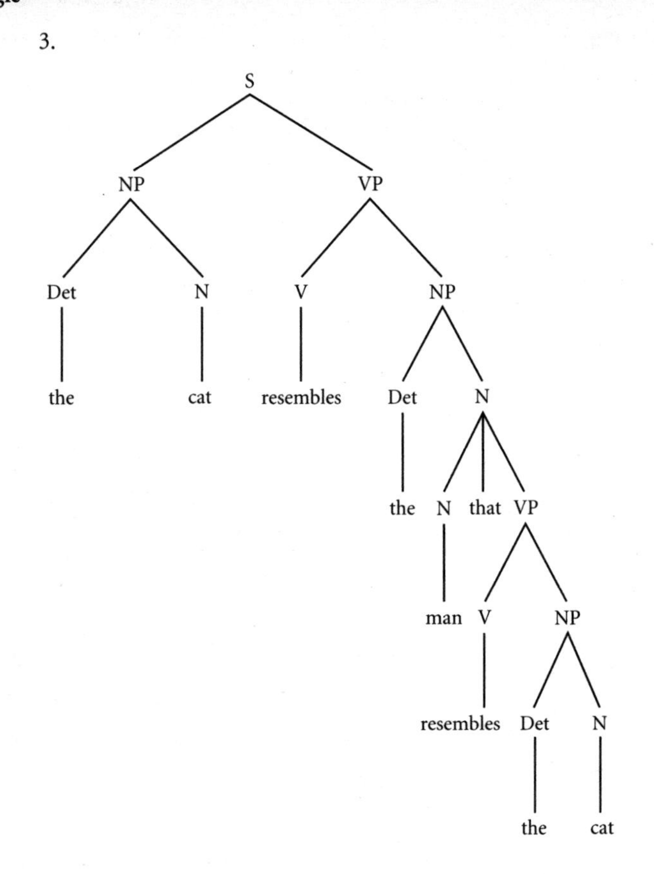
S
NP
VP
Det
N
V
NP
the
cat
resembles
Det
N
the
N
that
VP
man
V
NP
resembles
Det
N
the
cat

4.

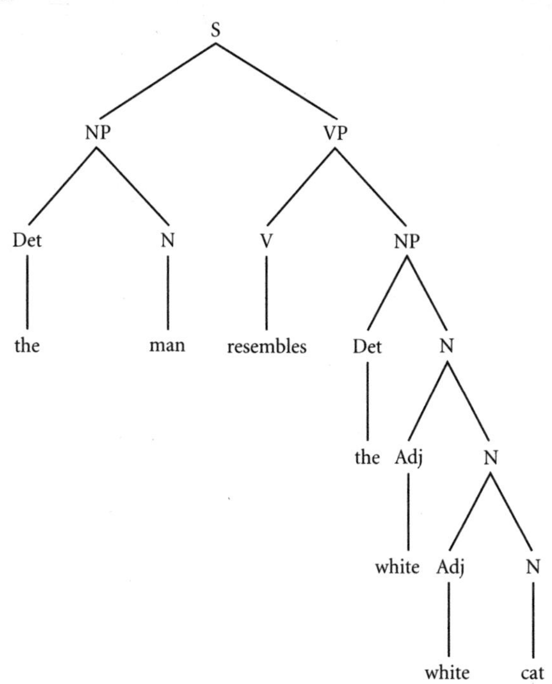
S
NP
VP
Det
N
V
NP
the
man
resembles
Det
N
the
Adj
N
white
Adj
N
white
cat

5.

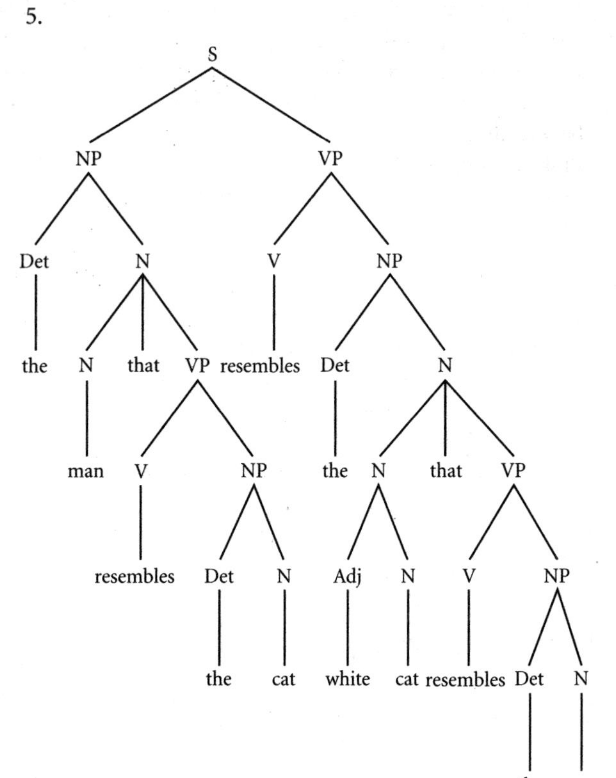

15C. S $\Rightarrow$ O
S $\Rightarrow$ LP O
S $\Rightarrow$ O RP
O $\Rightarrow$ boy
O $\Rightarrow$ sock
O $\Rightarrow$ mommy
LP $\Rightarrow$ allgone
LP $\Rightarrow$ byebye
RP $\Rightarrow$ off
RP $\Rightarrow$ on
RP $\Rightarrow$ fall

I took this example from P. S. Dale, *Language Development*, Dryden Press, Hinsdale, Ill., 1972, p. 41. As Dale explains, it used to be thought that children all over the world pass through a stage of using a simple grammar like that above. Today psychologists believe that the truth is much more complicated; see Stephen Pinker, 'Language Acquisition' in D. N. Osherson ed., *An Invitation to Cognitive Science, Vol. I, Language*, MIT Press, Cambridge, Mass., 1995, pp. 135–82.

Section 16

16A. ϕ_1, and then at once ϕ_2, so that ϕ_3; ϕ_4.
I scattered the strong warriors of Hadadezer.
I pushed the remnants of his troops into the Orontes.
They dispersed to save their lives.
Hadadezer himself perished.

16B. 1.

ϕ	It's a lie that ϕ
T	F
F	–

2.

ϕ	ψ	ϕ because ψ.
T	T	–
T	F	F
F	T	F
F	F	F

3.

ϕ	ψ	ϕ whenever ψ.
T	T	–
T	F	–
F	T	F
F	F	–

(The truth-values at just one time of the sentences put for 'ϕ' and 'ψ' can never suffice to make the whole sentence true, since the whole sentence talks about all the times at which the second constituent is true.)

4.

ϕ	If ϕ then ϕ.
T	T
F	T

5.

ϕ	Whether or not ϕ, what will be will be.
T	T
F	T

6.

ϕ	Whether or not ϕ, smoking causes cancer.
T	–
F	–

(The truth-value in 6 is the same as that of 'Smoking causes cancer', which is true in some possible situations and false in others.)

16C. 1. I am aware that [you intend to sue].
2. He regrets that [he didn't marry Suzy].
3. [He completed his task] before [the week ended].
4. [The train had an accident] because [its brakes failed].
5. *One possible answer:* Your Majesty may be pleased to notice that [great mischiefs may fall upon this kingdom] if [the intentions of bringing in Irish and foreign forces shall take effect]; it has been credibly reported that [there are intentions of bringing in Irish and foreign forces].

Section 17

17.1. ¬ Some dogs will be admitted. (NOT '¬ Dogs will be admitted'; think of the case where some dogs will be admitted and others won't.)
2. [the brain is bisected ∧ the character remains intact]
3. [the safety conditions will be tightened ∨ there is going to be a nasty accident]; *or* [there is going to be a nasty accident ∨ the safety conditions will be tightened]
4. [you're right → I stand to lose a lot of money]

5. [you broke the law ↔ the agreement formed a contract]
6. [somebody will call → I shall pretend I am designing St Paul's]
7. [Schubert is terrific ∧ Elvis Costello is terrific]. ('So' carries a cross-reference.)
8. [this is Bert Bogg ∧ Bert Bogg taught me that limerick I was quoting yesterday]
9. [you can claim the allowance → you earn less than £160 a week]. ('↔' would be wrong; there may be other conditions you have to satisfy besides earning less than £160 a week.)
10. [Liszt is horrible ∧ Vivaldi is horrible]
11. [she needs all the help she can get ∧ she's a single parent]
12. [the elder son was highly intelligent ∧ the younger son had learning difficulties]. (The second underlying constituent sentence has 'younger son'.)
13. ¬ Her performance had zest.
14. [he'll get something right → he'll pass]
15. [the metal will stretch ∨ the metal will snap]. (Another cross-reference.)

Section 18

18A. 1. [the nappies are becoming hard → you can soften the nappies by using a water conditioner]
2. *Impossible.*
3. *Impossible.*
4. [most of his former protection has worn off → his new vaccination develops much like the previous one]
5. *Impossible.*
6. [your baby is colicky → your baby may be soothed when you first pick him up]

18B. 1. *Impossible.*
2. *Impossible.*
3. [the evening was thoroughly pleasant ∧ the evening concluded with a waltz]
4. *Impossible.*
5. [Matilda's aunt had from earliest youth kept a strict regard for Truth ∧ Matilda's aunt attempted to believe her]

6. *Impossible.*
7. [Marianne is a teacher ∧ Marianne should have known better]
8. [Britain was once a superpower ∧ Britain is now seeking a new role]
9. *Impossible.* (Don might grow into a spectacle-free pedant, then into a bespectacled non-pedant.)

Section 19

1. ¬ ¬ I shall write to you. *or just* I shall write to you.
2. [he was gassed ∧ ¬ he was shot]
3. [¬ he was gassed ∧ ¬ he was shot] *or* ¬ [he was gassed ∨ he was shot]
4. [Tracey will scream again → ¬ somebody will get some chocolate]
5. [the private schools provide an excellent education ∧ [¬ the State schools have adequate space → the private schools ease the burden on the State's facilities]]
6. [[the private schools are socially top-heavy → the private schools are perpetuating social injustice] ∧ [the private schools are perpetuating social injustice → ¬ the private schools can reasonably demand charitable status]]

Section 20

20A. See p. 281–2 for the tableau derivation rules.

20B. 1. *Set of sentences:*

Mr Zak is a Russian spy. ¬[Mr Zak is a C.I.A. spy ∧ Mr Zak is a Russian spy]. [Mr Zak is a C.I.A. spy ∧ Mr Zak is a cad]

Tableau:

Mr Zak is a Russian spy.
✓ ¬[Mr Zak is a C.I.A. spy ∧
Mr Zak is a Russian spy]

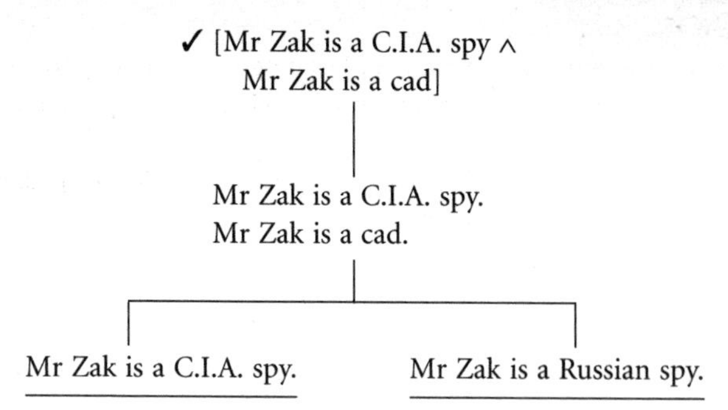
✓ [Mr Zak is a C.I.A. spy ∧
Mr Zak is a cad]
Mr Zak is a C.I.A. spy.
Mr Zak is a cad.
Mr Zak is a C.I.A. spy.
Mr Zak is a Russian spy.

Inconsistent.

2. *Set of sentences:*
   [Auguste lives in Bootle ∨ Bruno lives in Bootle]
   [Bruno is an estate agent ∨ Chaim is an estate agent]
   [¬ Bruno is an estate agent ∧ ¬ Bruno lives in Bootle]

   Tableau:

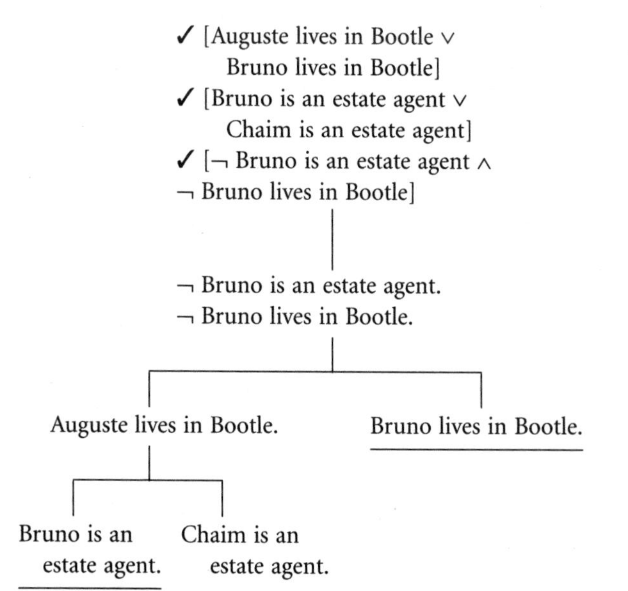
✓ [Auguste lives in Bootle ∨
Bruno lives in Bootle]
✓ [Bruno is an estate agent ∨
Chaim is an estate agent]
✓ [¬ Bruno is an estate agent ∧
¬ Bruno lives in Bootle]
¬ Bruno is an estate agent.
¬ Bruno lives in Bootle.
Auguste lives in Bootle.
Bruno lives in Bootle.
Bruno is an
estate agent.
Chaim is an
estate agent.

Consistent: middle branch won't close.

3. *Set of sentences:*

   [Yvonne gave me this book last Tuesday ∨ Zoe gave me this book last Tuesday]. [Yvonne gave me this book → I was in Oslo at Tuesday lunchtime]. [I was miles away from Oslo all Tuesday ∧ Zoe has never given me anything]

Tableau:

✓ [Yvonne gave me this book last Tuesday ∨ Zoe gave me this book last Tuesday]
✓ [Yvonne gave me this book → I was in Oslo at Tuesday lunchtime]
✓ [I was miles away from Oslo all Tuesday ∧ Zoe has never given me anything]

I was miles away from Oslo all Tuesday.
Zoe has never given me anything.

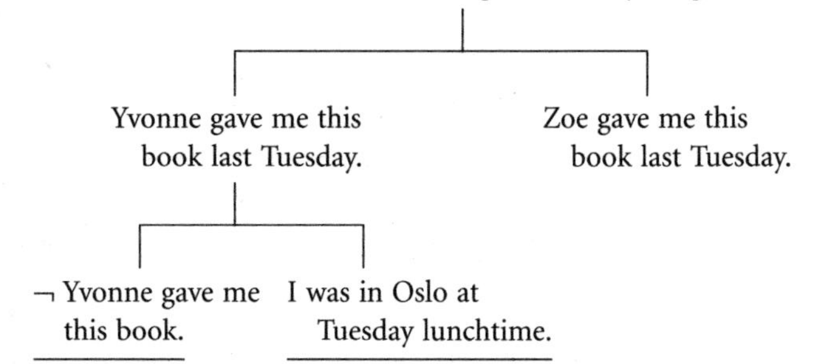
Yvonne gave me this book last Tuesday.
Zoe gave me this book last Tuesday.
¬ Yvonne gave me this book.
I was in Oslo at Tuesday lunchtime.

Inconsistent.

20C. 1. [the gunmen are tired → the gunmen are on edge]. [[the gunmen are armed ∧ the gunmen are on edge] → the hostages are in danger]. [the gunmen are armed ∧ the gunmen are tired]. *Therefore* the hostages are in danger.

2. [the driver was in control → [the driver passed the signal ↔ the signal was green]]. [the driver passed the signal ∧ the driver was in control]. [the signal was green → the electronics were faulty]. *Therefore* the electronics were faulty.

3. [the boy has spots in his mouth → the boy has measles]. [the boy has a rash on his back → the boy has heat-spots]. The boy has a rash on his back. *Therefore* ¬ the boy has measles.
4. [the vicar shot the earl ∨ the butler shot the earl]. [the butler shot the earl → ¬ the butler was drunk at nine o'clock]. [the vicar is a liar ∨ the butler was drunk at nine o'clock]. *Therefore* [the vicar is a liar ∨ the vicar shot the earl].

20D. 1.

✓ [the gunmen are tired →
the gunmen are on edge]
✓ [[the gunmen are armed ∧ the
gunmen are on edge] → the
hostages are in danger]
✓ [the gunmen are armed ∧
the gunmen are tired]
¬ the hostages are in danger.

The gunmen are armed.
The gunmen are tired.

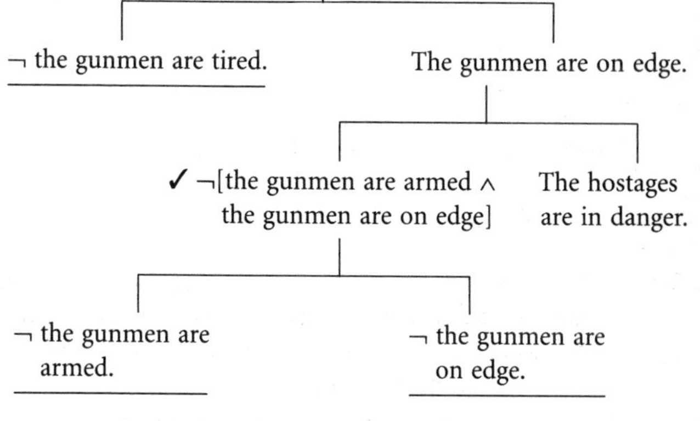
¬ the gunmen are tired.
The gunmen are on edge.
✓ ¬[the gunmen are armed ∧ the gunmen are on edge]
The hostages are in danger.
¬ the gunmen are armed.
¬ the gunmen are on edge.

Valid.

2.

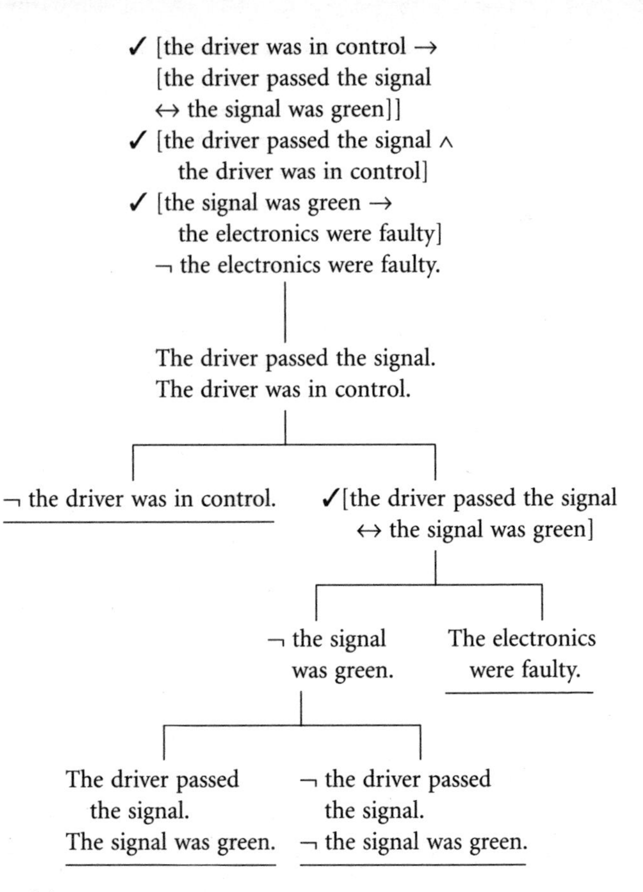
✓ [the driver was in control →
[the driver passed the signal
↔ the signal was green]]
✓ [the driver passed the signal ∧
the driver was in control]
✓ [the signal was green →
the electronics were faulty]
¬ the electronics were faulty.
The driver passed the signal.
The driver was in control.
¬ the driver was in control.
✓[the driver passed the signal
↔ the signal was green]
¬ the signal
was green.
The electronics
were faulty.
The driver passed
the signal.
The signal was green.
¬ the driver passed
the signal.
¬ the signal was green.

Valid.

3.

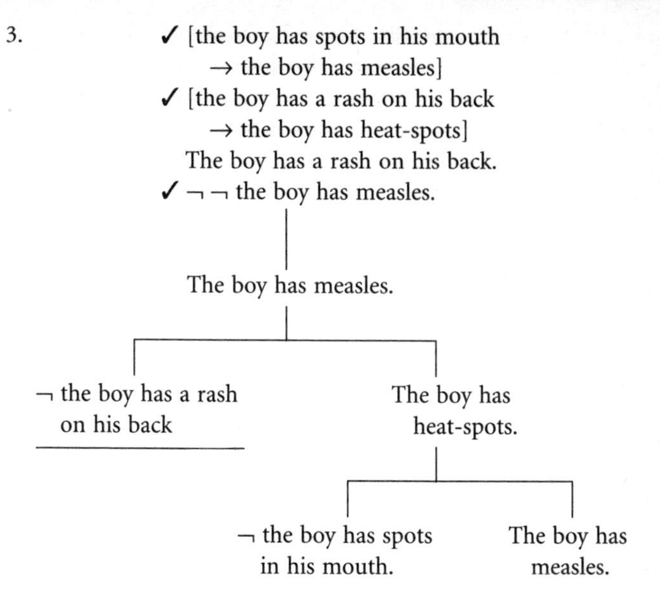

Invalid. A counterexample is the situation where the boy has both measles and heat-spots, and a rash on his back (see the right-hand branch).

4.

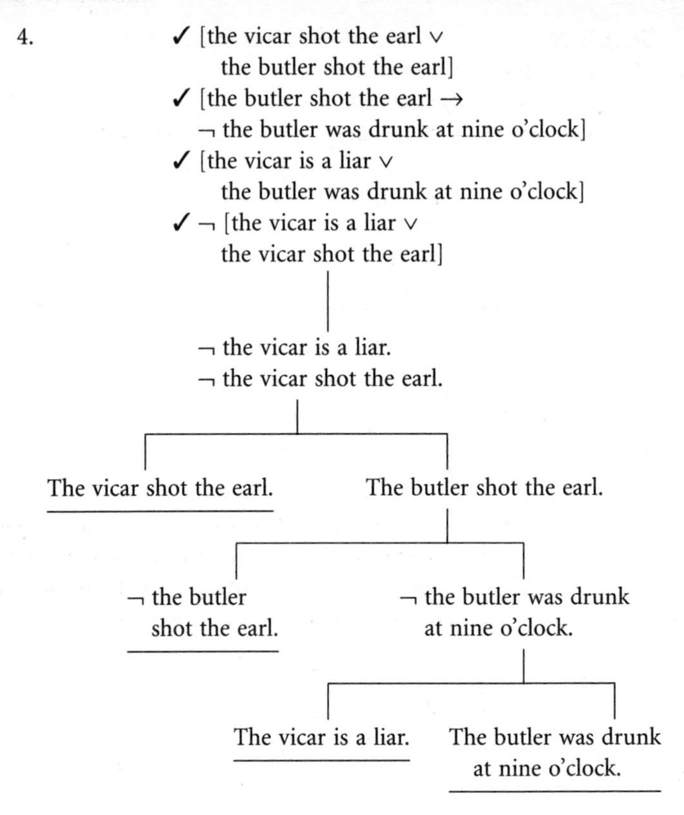
✓ [the vicar shot the earl ∨
the butler shot the earl]
✓ [the butler shot the earl →
¬ the butler was drunk at nine o'clock]
✓ [the vicar is a liar ∨
the butler was drunk at nine o'clock]
✓ ¬ [the vicar is a liar ∨
the vicar shot the earl]
¬ the vicar is a liar.
¬ the vicar shot the earl.
The vicar shot the earl.
The butler shot the earl.
¬ the butler
shot the earl.
¬ the butler was drunk
at nine o'clock.
The vicar is a liar.
The butler was drunk
at nine o'clock.

Valid.

Section 21

21A. 1. $\neg [Q \wedge R]$. $[P \rightarrow Q]$. R.

2. $[T \leftrightarrow R]$. $[Q \vee S]$. $[\neg Q \wedge T]$.

3. $[[Q \vee S] \rightarrow P]$. $[R \rightarrow T]$. $[[R \vee S] \wedge \neg T]$. $\neg P$.

21B. 1.

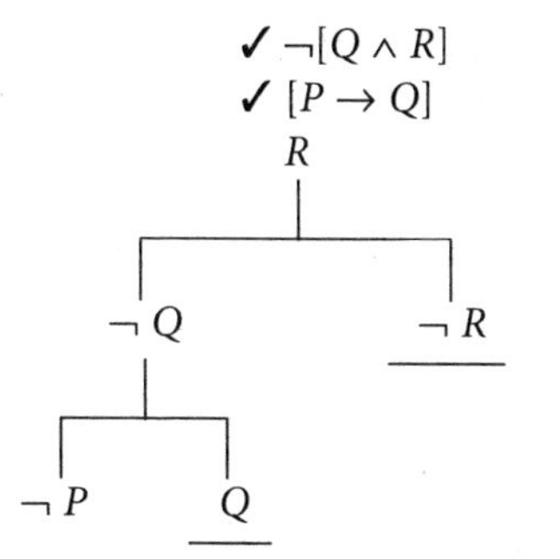

The left-hand branch describes a situation in which income derives from a dividend liable to capital gains tax and not from employment, and income tax is not levied on this income. There is no contradiction here; so the set is *consistent*.

2.

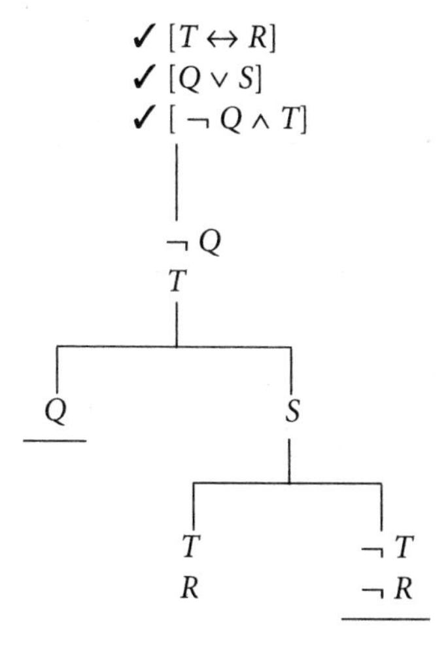

The central branch fails to close; but it describes a situation in which the source of this income is a dividend which both is and is not liable to capital gains tax – impossible. The other two branches are closed. Therefore the set is *inconsistent.*

3.

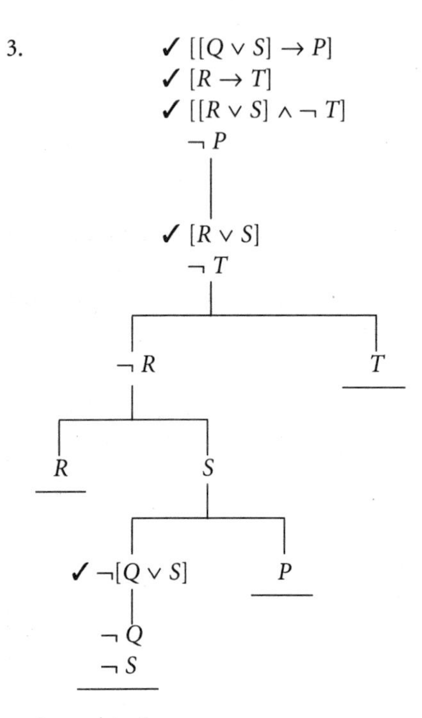

Inconsistent.

21C. 1. $[P \rightarrow T]$. $\neg P$. *Therefore* $\neg T$.

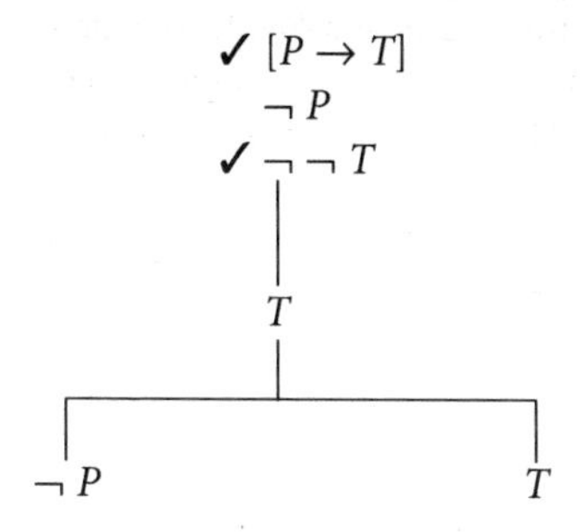

Invalid. The left branch gives the counterexample: Uhha-muwas has not bitten off Pissuwattis' nose, but is liable to a 30 shekel fine (say, for a traffic offence). The right branch also gives the same counterexample.

2. $[P \rightarrow [Q \vee S]]$. $\neg Q$. *Therefore* $[S \vee \neg P]$.

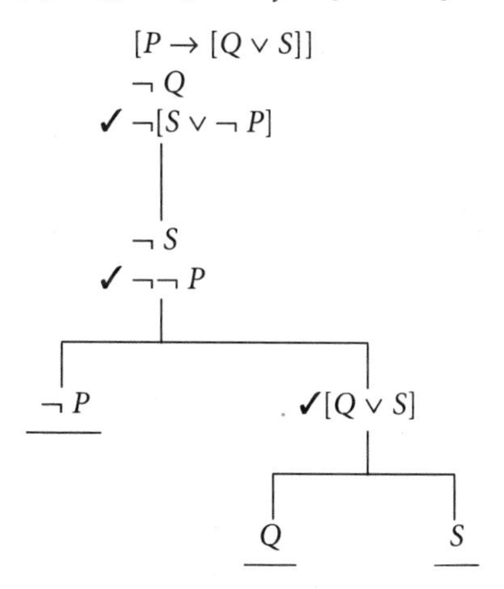

Valid. Tableau closed.

3. $[[S \rightarrow R] \wedge P]$. Q. *Therefore* $[\neg R \wedge [\neg S \wedge P]]$.

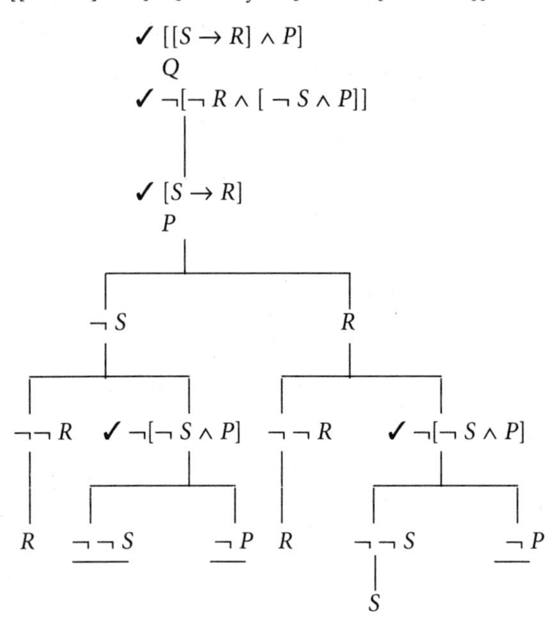

Valid. The tableau is not closed, but each of the unclosed branches describes a situation where Pissuwattis is both a female slave (Q) and a free woman (R), which is impossible.

4. $[P \rightarrow [[R \rightarrow S] \wedge [Q \rightarrow T]]]$. $[P \wedge \neg T]$. $[R \vee Q]$.
Therefore S. (Note that 'or' in the first sentence means '$\wedge$'.)

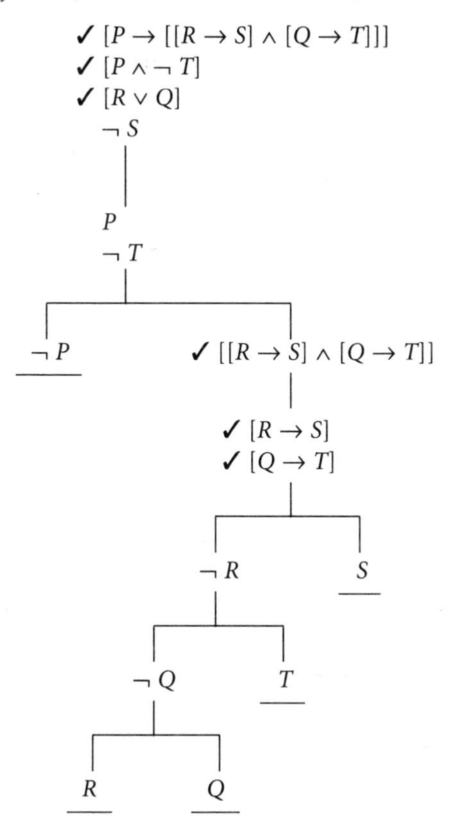

Valid. Tableau closed.

According to the laws of the ancient Hittites, 'If anyone bites off a free man's nose, he shall give 1 mina of silver and pledge his estate as security. If anyone bites off the nose of a male or female slave, he shall give 30 shekels of silver and pledge his estate as security.' (p. 189 of *Ancient Near Eastern Texts*, ed., J.B. Pritchard, Princeton University Press, 1955.)

Section 22

22.1. *Formula.*

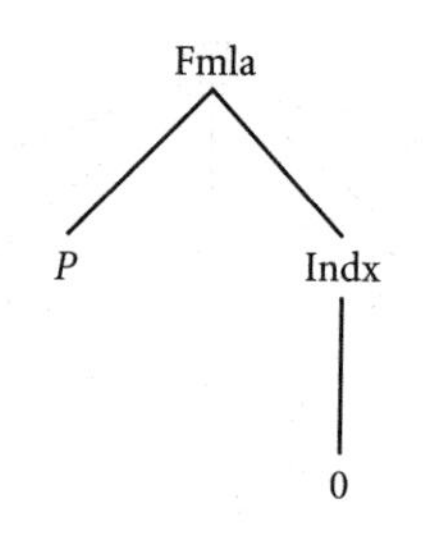

The only subformula is 'P_0' itself.

2. *Formula.*

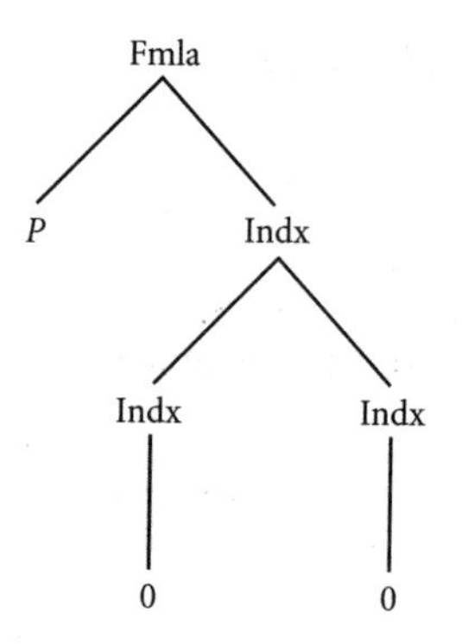

The only subformula is 'P_{00}' itself. ('P_0' is a formula but not a constituent of 'P_{00}', as the phrase-marker shows.)

3. *Formula.*

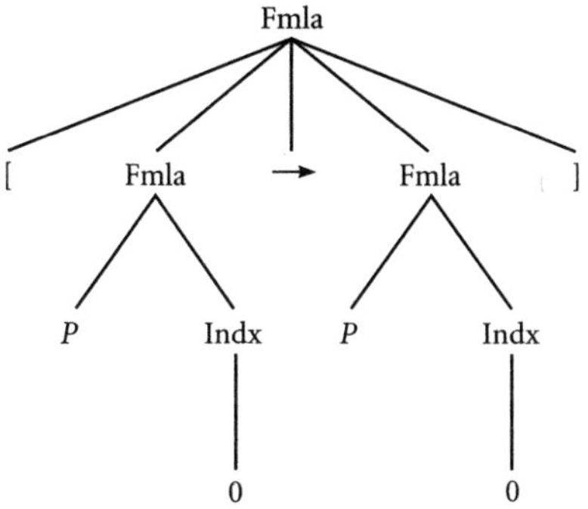

There are three subformulae:

$$\begin{array}{ll} & [P_0 \to P_0] \\ (1) & \underline{\hphantom{[P_0 \to P_0]}} \\ (2) & \hphantom{[}\underline{\hphantom{P_0}} \\ (3) & \hphantom{[P_0 \to{}}\underline{\hphantom{P_0}} \end{array}$$

4. *Not a formula.*
5. *Not a formula* (needs more brackets).
6. *Not a formula* (shouldn't have brackets).
7. *Not a formula.*

8. *Formula.*

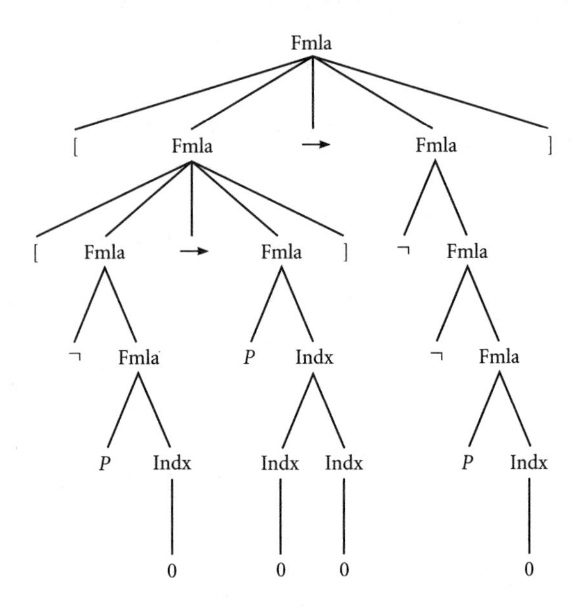

There are eight subformulae:

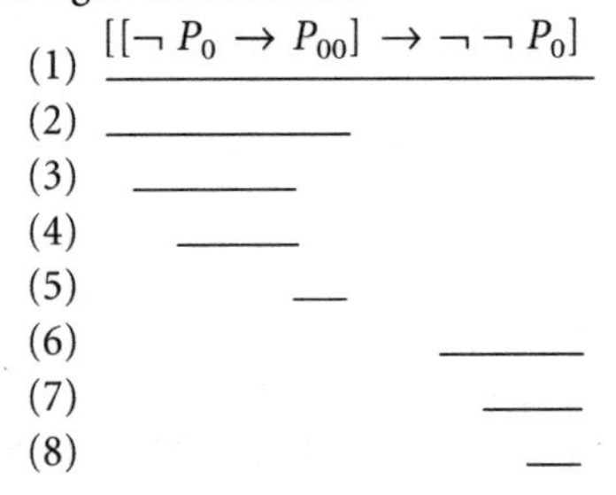

Section 23

23A. 1.

P	$[P \vee \neg P]$
T	T **T** F T
F	F **T** T F

2.

P	Q	$[[P \wedge Q] \vee [P \wedge \neg Q]]$
T	T	T T T **T** T F F T
T	F	T F F **T** T T T F
F	T	F F T **F** F F F T
F	F	F F F **F** F F T F

3.

P	Q	$[[P \rightarrow Q] \rightarrow [P \leftrightarrow Q]]$
T	T	T T T **T** T T T
T	F	T F F **T** T F F
F	T	F T T **F** F F T
F	F	F T F **T** F T F

4.

P	Q	$[[[P \rightarrow Q] \rightarrow P] \rightarrow P]$
T	T	T T T **T** T T T
T	F	T F F **T** T T T
F	T	F T T **F** F T F
F	F	F T F **F** F T F

5.

P	Q	$[[P \wedge Q] \wedge [\neg P \vee \neg Q]]$
T	T	T T T **F** F T F F T
T	F	T F F **F** F T T T F
F	T	F F T **F** T F T F T
F	F	F F F **F** T F T T F

6.

P	Q	R	$[[[P \rightarrow Q] \wedge [Q \rightarrow R]] \rightarrow [P \rightarrow R]]$
T	T	T	T T T T T T T **T** T T T
T	T	F	T T T F T F F **T** T F F
T	F	T	T F F F F T T **T** T T T
T	F	F	T F F F F T F **T** T F F
F	T	T	F T T T T T T **T** F T T
F	T	F	F T T F T F F **T** F T F
F	F	T	F T F T F T T **T** F T T
F	F	F	F T F T F T F **T** F T F

23B. 1, 4 and 6 are tautologies; the rest are not.

23C. 1.

P	Q	P. $[P \rightarrow Q] \models$	
T	T	T	T
T	F	T	F
F	T	F	T
F	F	F	T

Incorrect. Counterexample

P	Q
T	T

2.

P	Q	$P \models [Q \rightarrow P]$	
T	T	T	T
T	F	T	T
F	T	F	F
F	F	F	T

Correct.

3.

P	Q	P. $\neg P \models Q$		
T	T	T	F	T
T	F	T	F	F
F	T	F	T	T
F	F	F	T	F

Correct: from a contradiction anything follows.

4.

P	Q	$[Q \rightarrow P]$.	$[Q \rightarrow \neg P] \models$	$\neg Q$
T	T	T	F	F
T	F	T	T	T
F	T	F	T	F
F	F	T	T	T

Correct.

5.

P	Q	R	$[P \to Q]$.	$[Q \to R] \models$	$[R \to P]$
T	T	T	T	T	T
T	T	F	T	F	T
T	F	T	F	T	T
T	F	F	F	T	T
F	T	T	T	T	F
F	T	F	T	F	T
F	F	T	T	T	F
F	F	F	T	T	T

Incorrect. Counterexample

P	Q	R
F	T	T

6.

P	Q	R	$[R \leftrightarrow [P \vee Q]] \models$	$[[R \wedge \neg P] \to Q]$
T	T	T	T	T
T	T	F	F	T
T	F	T	T	T
T	F	F	F	T
F	T	T	T	T
F	T	F	F	T
F	F	T	F	F
F	F	F	T	T

Correct.

Section 24

24A. 1. Tautology 2. $P \mapsto [Q \to R]$

2. Tautology 33. $Q \mapsto [P \to P]$

3. Tautology 35. $Q \mapsto [P \to P]$
$R \mapsto P$

4. Tautology 26. $P \mapsto [Q \leftrightarrow R]$
$Q \mapsto P$

24B. 1. $\neg [\neg [P \wedge \neg Q] \wedge \neg [\neg P \wedge Q]]$

2. $\neg [[\neg P \wedge \neg Q] \wedge \neg R]$

3. $[\neg [\neg P \wedge \neg Q] \wedge \neg P]$

24C. $[[[[\phi \wedge \psi] \wedge \chi] \vee [[\phi \wedge \neg \psi] \wedge \neg \chi]] \vee [[\neg \phi \wedge \psi] \wedge \chi]]$

24D. '$\neg R$' is the simplest interpolant. The method described in the text would lead to the interpolant '$[[Q \wedge \neg R] \vee [\neg Q \wedge \neg R]]$'.

Section 25

25. Tautology 9.

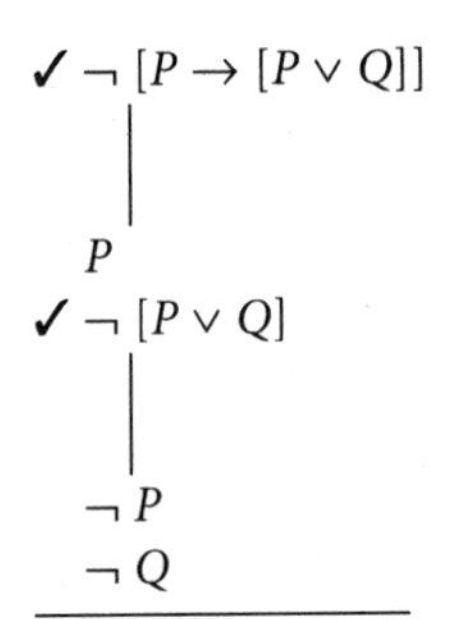

Tautology 11.

✓ $\neg [[P \to R] \to [[Q \to R] \to [[P \vee Q] \to R]]]$

✓ $[P \to R]$

✓ $\neg [[Q \to R] \to [[P \vee Q] \to R]]$

✓ $[Q \to R]$

✓ $\neg [[P \vee Q] \to R]$

✓ $[P \vee Q]$

$\neg R$

$\neg P$ | R

$\neg Q$ | R

P | Q

Tautology 16.

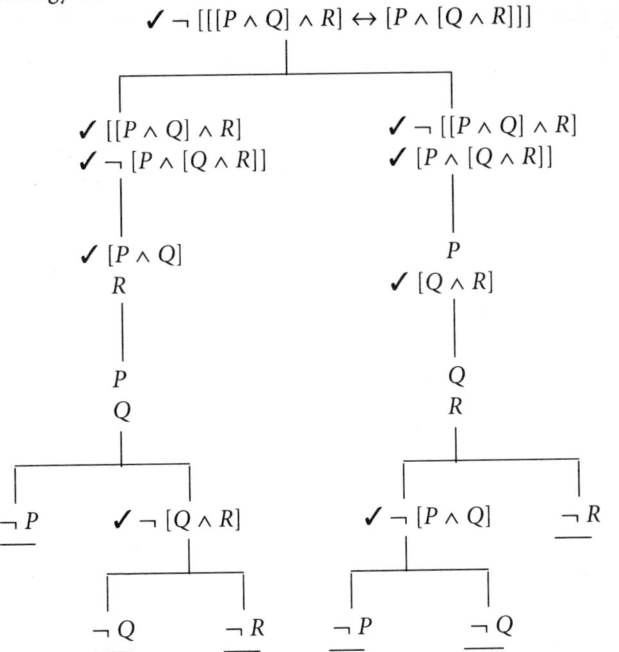

Tautology 26.

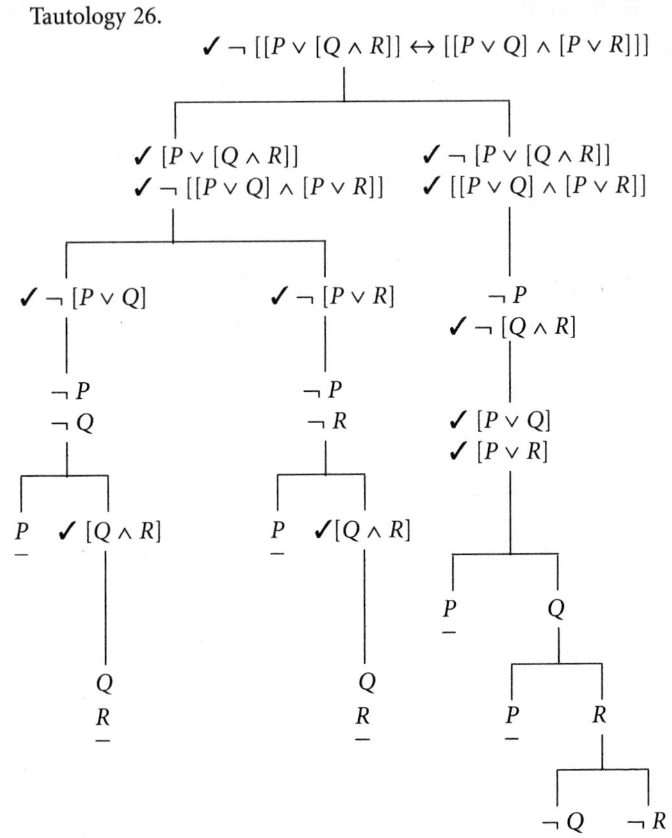

Tautology 48.

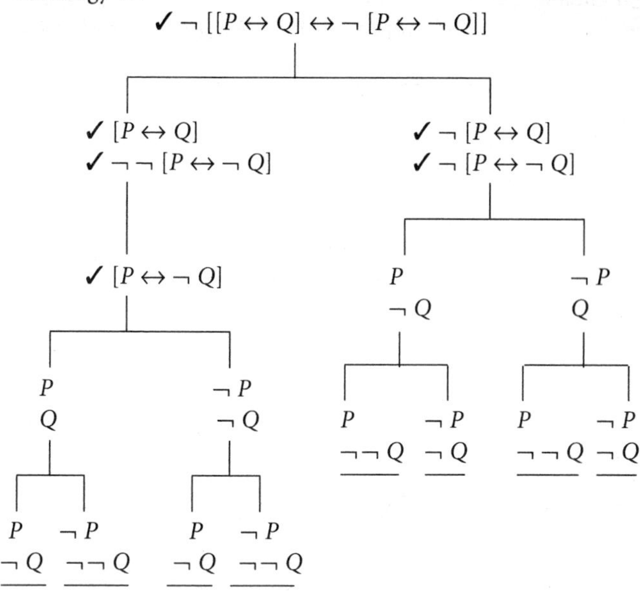

Section 26

26\. 1. I (singular personal pronoun); he (spp); this (definite description); the ox (dd); the day (dd); relief (non-count noun); its back (dd); I (spp); he (spp); he (spp); God (proper name). If 'its back being rubbed' is counted as a singular noun phrase, then this is a definite description too.

2. this rule (dd); Dr Jekyll (pn); he (spp); the opposite side of the fire (dd); the opposite side (dd); the fire (dd); capacity (ncn); kindness (ncn); you (spp); he (spp); Mr Utterson (pn). Although **affection** is normally a non-count noun, the 'a' in front of it shows that it is not a non-count noun here.

Section 27

27\. 1. No: intentional.

2. Yes. (Remember that our paraphrase is allowed to assume that she does have a lover.)

3. No: quotational.

4. No: modal.

5. No: space-time.
6. Yes.
7. Yes.
8. No: quotational.
9. No: intentional.
10. Yes (in spite of the word 'impossible').
11. No: space-time (in spite of the intentional phrase 'nagging feeling'); it's the oven at the house, not necessarily the oven here.
12. First occurrence, No: space-time. Second occurrence, presumably No by our test, since any sentence along the lines of 'The king is a person such that long live him!' is going to be ungrammatical; but in fact the reference of the phrase here does seem to be its primary reference.

Section 29

29\. 1. Three plus three is greater than five.
2. Thy father's spirit scents the morning air. (Change 'scent' to 'scents', to correct the perturbation.)
3. Parasurama is Krishna.
4. (i) Parasurama is Krishna. (ii) Krishna is Parasurama.
5. (i) The daughters of the king disown Lear. (ii) The daughters of Lear disown the king. (iii) The daughters of the king disown the king.
6. (i) Errol owns the gun that fired the bullet that killed Cheryl. (ii) The gun that fired this bullet is the murder weapon. (iii) (Using Leibniz's Rule twice) Errol owns the murder weapon.
7. (i) June is the third month after the ninth month after June. (ii) March is the ninth month after the third month after March. (iii) June is the third month after the ninth month after the third month after March. (etc., etc.)
8. The Father suffers. (This is a time-honoured problem for Christian theologians, who must accept the premises but deny the conclusion, on pain of heresy.)

Section 30

30A. 1. U.S.S.R. 2. U.S.S.R., Sweden. 3. Sweden.

30B. ⟨U.S.A., U.S.A.⟩, ⟨U.S.A., U.S.S.R.⟩, ⟨U.S.A., Sweden⟩, ⟨U.S.S.R., U.S.A.⟩, ⟨U.S.S.R., U.S.S.R.⟩, ⟨U.S.S.R., Sweden⟩, ⟨Sweden, U.S.A.⟩, ⟨Sweden, U.S.S.R.⟩, ⟨Sweden, Sweden⟩.

30C. 1. ⟨U.S.S.R., Sweden⟩, ⟨U.S.S.R., U.S.A.⟩, ⟨Sweden, U.S.A.⟩.
2. ⟨U.S.A., U.S.A.⟩, ⟨U.S.A., U.S.S.R.⟩, ⟨U.S.A., Sweden⟩, ⟨U.S.S.R., U.S.S.R.⟩, ⟨Sweden, Sweden⟩, ⟨Sweden, U.S.S.R.⟩.
3. ⟨U.S.A., U.S.A.⟩, ⟨U.S.A., U.S.S.R.⟩, ⟨U.S.A., Sweden⟩.

30D. {⟨1, 1⟩, ⟨1, 2⟩, ⟨2, 1⟩, ⟨2, 2⟩}, {⟨1, 1⟩, ⟨1, 2⟩, ⟨2, 1⟩}, {⟨1, 1⟩, ⟨1, 2⟩, ⟨2, 2⟩}, {⟨1, 1⟩}, ⟨2, 1⟩, ⟨2, 2⟩}, {⟨1, 2⟩, ⟨2, 1⟩, ⟨2, 2⟩}, {⟨1, 1⟩, ⟨1, 2⟩}, {⟨1, 1⟩, ⟨2, 1⟩}, {⟨1, 1⟩, ⟨2, 2⟩}, {⟨1, 2⟩, ⟨2, 1⟩}, {⟨1, 2⟩, ⟨2, 2⟩}, {⟨2, 1⟩, ⟨2, 2⟩}, {⟨1, 1⟩}, {⟨1, 2⟩}, {⟨2, 1⟩}, {⟨2, 2⟩}, {}.

Section 31

31A. 1. reflexive. 2. irreflexive. 3. non-reflexive. 4. irreflexive. 5. reflexive. 6. non-reflexive.

31B. 1. asymmetric. 2. symmetric. 3. asymmetric. 4. non-symmetric. 5. symmetric. 6. non-symmetric.

31C. 1. non-transitive. 2. transitive. 3. intransitive. 4. transitive. 5. intransitive, if people married to each other must be of opposite sex. 6. intransitive or non-transitive; the latter if the world contains an Oedipus (who married his mother, and had the daughter Antigone by her).

31D. 1. not connected. 2. connected. 3. connected. 4. not connected.

Section 32

32A. P1 has the same empirical formula as P1.
P2 has the same empirical formula as P2.
M has the same empirical formula as M.
P2 has the same empirical formula as P1.
M has the same empirical formula as P2.
P1 has the same empirical formula as M.
M has the same empirical formula as P1.

32B. 1. x_1 is the same size as x_2.
2. x_1 is the same age as x_2.
3. x_1 is the same distance as x_2.
4. x_1 does the same amount as x_2.
5. x_1 has the same frequency of breaking (or breakage rate) as x_2.
6. *See section 33.*

32C. 1. x_1 is greedier than x_2.
2. x_1 has fewer antennae than x_2.

3. x_1 is at least as knobbly as x_2.
4. x_1 has the same temperature as x_2.

Section 34

34. 1. *First group*: A man came to see me this afternoon.
    Second group: A gentleman doesn't pick his nose.
 2. *First group*: Drinks will be provided.
    Second group: Cats have an acute sense of smell.

Section 35

35. 1. $\forall x$ [x is a noise $\rightarrow$ x appals me]
 2. $\exists x$ [x is wicked $\wedge$ x comes this way]
 3. $\exists x$ [x is a strange infirmity $\wedge$ I have x]
 4. $\forall x$ [x is a candle of theirs $\rightarrow$ x is out]
 5. $\forall x$ x is not a child of his.
 6. $\exists x$ [x is a murder $\wedge$ x has been performed]
 7. $\exists y$ [y is an idiot $\wedge$ x is a tale told by y]
 8. $\forall y$ [y is a person born of woman $\rightarrow$ y shall not harm x]

Section 36

36A. [John is an older boy than John $\rightarrow$ John likes to sing].
[the boy in the corner is an older boy than John $\rightarrow$ the boy in the corner likes to sing].
[the corner is an older boy than John $\rightarrow$ the corner likes to sing].
(Presumably only the second of these sentences will be of any interest; but the first and third will be true too, since their first halves will be false.)

36B. 1. [the number ½ is odd $\vee$ the number ½ is even].
The number ½ is neither odd nor even.
 2. [Case 946 is a person $\rightarrow$ Case 946's I.Q. never varies by more than 15].
Case 946 is a girl whose I.Q. has varied between 142 and 87.
 3. [Horst is Austrian $\rightarrow$ Horst is of Alpine type].
[Horst is of Alpine type $\rightarrow$ Horst has a broad head].
Horst is Austrian, but doesn't have a broad head.
 4. [[the female cat over the road is a cat $\wedge$ the female cat over the road has two ginger parents] $\rightarrow$ the female cat over the road is ginger].

[the female cat over the road is a female cat → the female cat over the road is not ginger].
The female cat over the road has two ginger parents.

A tableau for 4:

✓ [[the female cat over the road is a cat ∧ the female cat over the road has two ginger parents] → the female cat over the road is ginger]
✓ [the female cat over the road is a female cat → the female cat over the road is not ginger]
The female cat over the road has two ginger parents.

¬ the female cat over the road is a female cat.
———

The female cat over the road is not ginger.

✓ ¬[the female cat over the road is a cat ∧ the female cat over the road has two ginger parents]

The female cat over the road is ginger.
———

¬ the female cat over the road is a cat.
———

¬ the female cat over the road has two ginger parents.
———

36C For example,
Maxine is not dead.
[the spouse of Maxine is a widow → the spouse of the spouse of Maxine is dead].
The spouse of the spouse of Maxine is not dead.

36D. 1. [*b* is a jockey ∧ *b* is female].
[*b* is female → *b* is not a jockey].

2. [*b* is male ∧ *b* does the cleaning].
[*b* does the cleaning → *b* doesn't go out to work].
[*b* is male → *b* goes out to work].

3. *b* has forty-seven chromosomes.
   [*b* is male ∨ *b* is female].
   [*b* is male → *b* has just forty-six chromosomes].
   [*b* is female → *b* has just forty-six chromosomes].
4. [*b* is male ∧ the mate of *b* is male].
   [*b* is male → the mate of *b* is female].
   [the mate of *b* is female → the mate of *b* is not male].

36E. 1. ¬ [*b* is a dunnock → *b* is not brightly coloured].
   [*b* is a ground-feeding bird → *b* is not brightly coloured].
   [*b* is a dunnock → *b* is a ground-feeding bird].
2. [*b* is a finch ∧ *b* cracks cherry seeds].
   [*b* is a finch → *b* is a bird].
   [[*b* is a bird ∧ *b* cracks cherry seeds] → *b* has a massive beak].
   ¬ [*b* is a finch ∧ *b* has a massive beak].
3. ¬ [*b* is one of the birds → *b* is a willow-warbler].
   [*b* is one of the birds → [*b* is a chiff-chaff ∨ *b* is a willow-warbler]].
   [*b* is one of the birds → *b* is singing near the ground].
   [*b* is a chiff-chaff → *b* doesn't sing near the ground].
4. *b* is a trumpeter bullfinch.
   [*b* is a trumpeter bullfinch → *b* can sing two notes at once].
   [*b* is a trumpeter bullfinch → *b* is a bird].
   ¬ [*b* is a bird ∧ *b* can sing two notes at once].

Section 37

37A. 1. ¬ ∀*x* [*x* is a time → the room is heated at *x*]
2. ∀*x* [*x* is a time → ¬ the room is heated at *x*]
3. As 2.
4. ∃*x* [*x* is a time ∧ ¬ the room is heated at *x*]
5. ¬ ∃*x* [*x* is a time ∧ the room is heated at *x*]
   (4 is a paraphrase of 1, and 5 is a paraphrase of 2.)

37B. As I read them, 1 and 2 imply one girl got the lot, 4–6 don't, and 3 is doubtful. (In revising for the second edition I wondered whether this exercise was politically correct. But I decided to retain it in memory of a much-missed friend, the late Sid Trivus of Los Angeles, who told me that in his high school days all the sentences of the exercise were absolutely true and it was the story of his life.)

37C. 1. $\forall x$ [[x is a starling $\wedge$ x is nesting here] $\rightarrow$ I'll shoot x]
2. $\forall x$[[x is a starling $\wedge$ x is nesting here] $\rightarrow$ that bird is a starling]
OR: [$\exists x$[x is a starling $\wedge$ x is nesting here] $\rightarrow$ that bird is a starling]
3. $\forall x$ [x is a finch $\rightarrow$ that bird has a longer bill than x]
4. $\exists x$ [x is a finch $\wedge \neg$ that bird has a longer bill than x]
5. $\forall x$ [x is a starling $\rightarrow \neg$ that bird has a longer bill than x]
6. $\neg \forall x$ [x is a starling $\rightarrow \forall y$ [y is a finch $\rightarrow$ x has a longer bill than y]]
7. $\exists x$ [x is a finch $\wedge \forall y$ [y is a starling $\rightarrow$ x has a longer bill than y]]
8. $\forall x$ [x is a finch $\rightarrow \exists y$ [y is a starling $\wedge$ y has a longer bill than x]]
9. $\forall x$ [[[x is a bird $\wedge$ x is nesting here] $\wedge \neg$ x is a finch] $\rightarrow$ x has a long bill]
10. $\forall x$ [[x is a bird $\wedge$ x is nesting here] $\rightarrow$ [x has a long bill $\leftrightarrow$ x is a starling]]

Section 38

38A. 1. $\exists x_1 \exists x_2$ [$\neg x_1 = x_2 \wedge$ [x_1 is a mistake $\wedge$ x_2 is a mistake]]
2. $\exists x_1 \exists x_2 \exists x_3 \exists x_4$ [[[[[[$\neg x_1 = x_2 \wedge \neg x_1 = x_3] \wedge \neg x_1 = x_4] \wedge \neg x_2 = x_3] \wedge \neg x_2 = x_4] \wedge \neg x_3 = x_4] \wedge$ [[[x_1 is a mistake $\wedge$ x_2 is a mistake] $\wedge$ x_3 is a mistake] $\wedge$ x_4 is a mistake]]
3. $\exists x_1 \exists x_2$ [$\neg x_1 = x_2 \wedge$ [[x_1 is a person $\wedge$ x_1 has pointed out the mistakes] $\wedge$ [x_2 is a person $\wedge$ x_2 has pointed out the mistakes]]]
4. [$\exists x_1 \exists x_2$ [$\neg x_1 = x_2 \wedge$ [x_1 is a hemisphere $\wedge$ x_2 is a hemisphere]] $\wedge \neg \exists x_1 \exists x_2 \exists x_3$ [[[$\neg x_1 = x_2 \wedge \neg x_1 = x_3] \wedge \neg x_2 = x_3] \wedge$ [[x_1 is a hemisphere $\wedge$ x_2 is a hemisphere] $\wedge$ x_3 is a hemisphere]]]
5. $\forall x$ [x = Sir Henry $\leftrightarrow$ x is allowed to use that bath]

38B. 1. [$\exists x \forall y$ [$x = y \leftrightarrow y$ is a book] $\wedge \forall x$ [x is a book $\rightarrow$ x is bound in vellum]]
2. [$\exists x \forall y$ [$x = y \leftrightarrow y$ is a stern of hers] $\wedge \forall x$ [x is a stern of hers $\rightarrow$ we've avoided x]]
3. Impossible – there may be other sons.
4. $\forall x$ [x = my mother $\leftrightarrow$ x is a woman in blue]

5. Impossible – not purely referential.
6. Impossible – non-count noun.
7. $\forall x\ [\neg\ x = \text{Cassius} \rightarrow \text{Cassius is greater than } x]$

Section 39

39A.

1. Ec
2. $\neg\ Wdb$
3. $[Sbc \vee Scb]$
4. $\exists x[[Ex \wedge Wdx] \wedge Sbx]$
5. $\exists x \exists y[\neg\ x = y \wedge [[Ex \wedge Wdx] \wedge [Ey \wedge Wdy]]]$
   (It seems to me that 5 implies that Homer wrote at least one epic, unlike '¬ Homer wrote just one epic'. See p. 71.)
6. $[Eb \wedge \exists x[Ex \wedge \neg\ Sbx]]$
7. $\forall x[Wxb \rightarrow \neg\ Wxc]$
8. $\forall x[[Ex \wedge Sxb] \rightarrow \neg\ Wdx]$
9. $\forall x[Ex \rightarrow \exists y[Ey \wedge [Wdy \wedge Sxy]]]$
10. $\exists x \exists y[[[[Ex \wedge Scx] \wedge [Ey \wedge Scy]] \wedge$
    $\forall z[[Ez \wedge Scz] \rightarrow [z = x \vee z = y]]] \wedge [Wdx \wedge \neg\ Wdy]]$

39B.

1. $\forall x \neg\ Rxx$
2. $\forall x \forall y[Rxy \rightarrow Ryx]$
3. $\forall x \forall y[Rxy \rightarrow \neg\ Ryx]$
4. $\forall x \forall y \forall z[[Rxy \wedge Ryz] \rightarrow \neg\ Rxz]$
5. $[\neg\ \forall x Rxx \wedge \neg\ \forall x \neg\ Rxx]$
6. $\forall x \forall y[\neg\ x = y \rightarrow [Rxy \wedge Ryx]]$

Section 40

40A. $[M \wedge \forall x[Gx \rightarrow [\neg Sx \vee [Cx \wedge Lx]]]]$. $[Gd \wedge [Cd \wedge \neg Ld]]$.
Therefore $\neg Sd$.

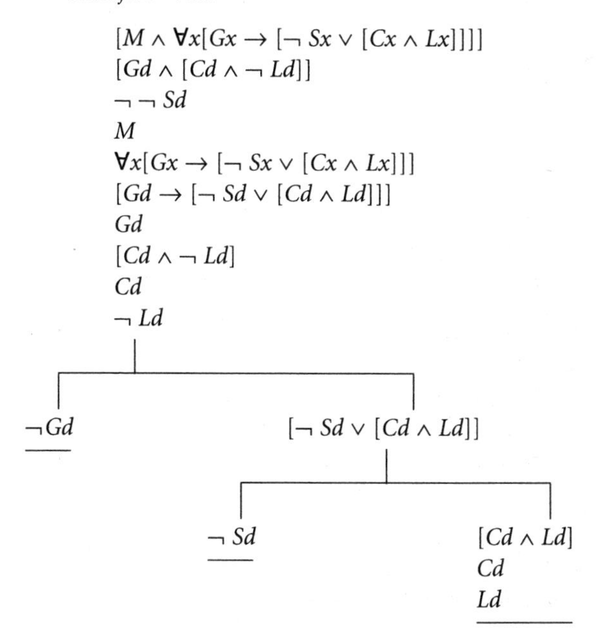

40B. $\forall x \forall y[Rxy \rightarrow \neg Ryx]$

$\neg \forall x \neg Rxx$

$\exists x \neg \neg Rxx$

$\neg \neg Rbb$

$\forall x[Rby \rightarrow \neg Ryb]$

$[Rbb \rightarrow \neg Rbb]$

$\neg Rbb$ | $\neg Rbb$

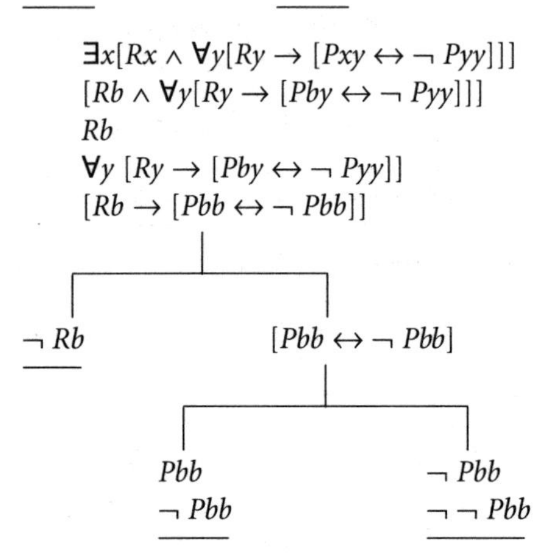

40D. 1. Pb

$\neg \forall x[x = b \rightarrow Px]$

$\exists x \neg [x = b \rightarrow Px]$

$\neg [c = b \rightarrow Pc]$

$c = b$

$\neg Pc$

$\neg Pb$

2.

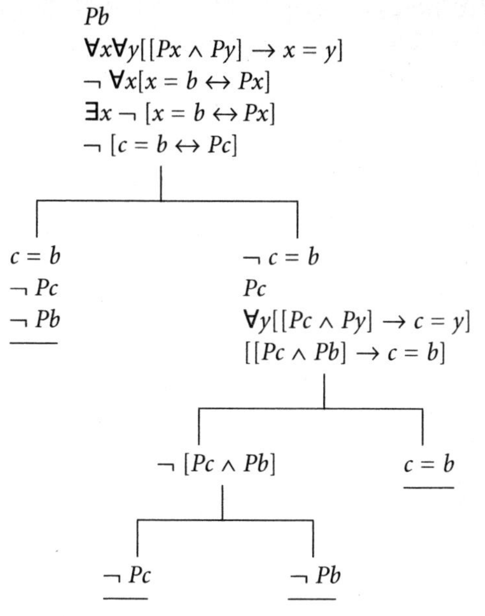
Pb
∀x∀y[[Px ∧ Py] → x = y]
¬ ∀x[x = b ↔ Px]
∃x ¬ [x = b ↔ Px]
¬ [c = b ↔ Pc]
c = b
¬ Pc
¬ Pb
¬ c = b
Pc
∀y[[Pc ∧ Py] → c = y]
[[Pc ∧ Pb] → c = b]
¬ [Pc ∧ Pb]
c = b
¬ Pc
¬ Pb

3.

$\forall x[x = b \leftrightarrow Px]$

$\neg [Pb \wedge \forall x \forall y [[Px \wedge Py] \rightarrow x = y]]$

$\neg Pb$

$[b = b \leftrightarrow Pb]$

$b = b$

Pb

$\neg b = b$

$\neg Pb$

$\neg \forall x \forall y [[Px \wedge Py] \rightarrow x = y]$

$\exists x \neg \forall y [[Px \wedge Py] \rightarrow x = y]$

$\neg \forall y [[Pc \wedge Py] \rightarrow c = y]$

$\exists y \neg [[Pc \wedge Py] \rightarrow c = y]$

$\neg [[Pc \wedge Pd] \rightarrow c = d]$

$[Pc \wedge Pd]$

$\neg c = d$

Pc

Pd

$[c = b \leftrightarrow Pd]$

$[d = b \leftrightarrow Pd]$

$c = b$

Pc

$d = b$

Pd

$c = d$

$\neg d = b$

$\neg Pd$

$\neg c = b$

$\neg Pc$

40E. 1.

$\forall x Px$

$\forall x Qx$

$\neg \exists x[Px \wedge Qx]$

$\forall x \neg [Px \wedge Qx]$

$\neg [Pa \wedge Qa]$ (by VII)

Pa

Qa

$\neg Pa$

$\neg Qa$

---

2. $[\forall xPx \vee \exists yQy]$
$\neg\exists y[\forall xPx \vee Qy]$
$\forall y\neg[\forall xPx \vee Qy]$

Left branch:
$\forall xPx$
$\neg[\forall xPx \vee Qc]$ (by VII)
$\neg\forall xPx$
$\neg Qc$

Right branch:
$\exists yQy$
Qa
$\neg[\forall xPx \vee Qa]$
$\neg\forall xPx$
$\neg Qa$

40F. The symbolizations first:

1. $\forall x[[Mx \wedge Cx] \rightarrow \forall y[Py \rightarrow Sxy]]$. $\exists xPx$.
$\forall x[[Cx \wedge \exists y[Py \wedge Sxy]] \rightarrow Bx]$. $[Mb \wedge Cb]$.
Therefore Bb.
2. $\forall x[Px \rightarrow [Sbx \leftrightarrow Scx]]$. $\forall x[[Px \wedge Sbx] \rightarrow Bb]$. $\neg Bb$.
Therefore $\forall x[Px \rightarrow \neg Scx]$.
3. $\exists x\exists y[[Cx \wedge Cy] \wedge [Txy \vee Tyx]]$. $\forall x[Cx \rightarrow \neg Txx]$.
Therefore $\exists x\exists y[\neg x = y \wedge [Cx \wedge Cy]]$.
4. $\neg b = c$. $[\neg Bb \vee \forall x[Px \rightarrow Sbx]]$.
$[Cc \wedge [Tcb \rightarrow \exists x[Px \wedge [Scx \wedge \neg Sbx]]]]$.
$\forall x[[Cx \wedge \neg b = x] \rightarrow Txb]$. *Therefore* $\neg Bb$.
5. $\forall x[x = b \leftrightarrow [Mx \wedge Cx]]$. $\forall x[Cx \rightarrow [Bx \leftrightarrow Mx]]$.
Therefore $\forall x[x = b \leftrightarrow [Cx \wedge Bx]]$.

Next the tableaux:

1.

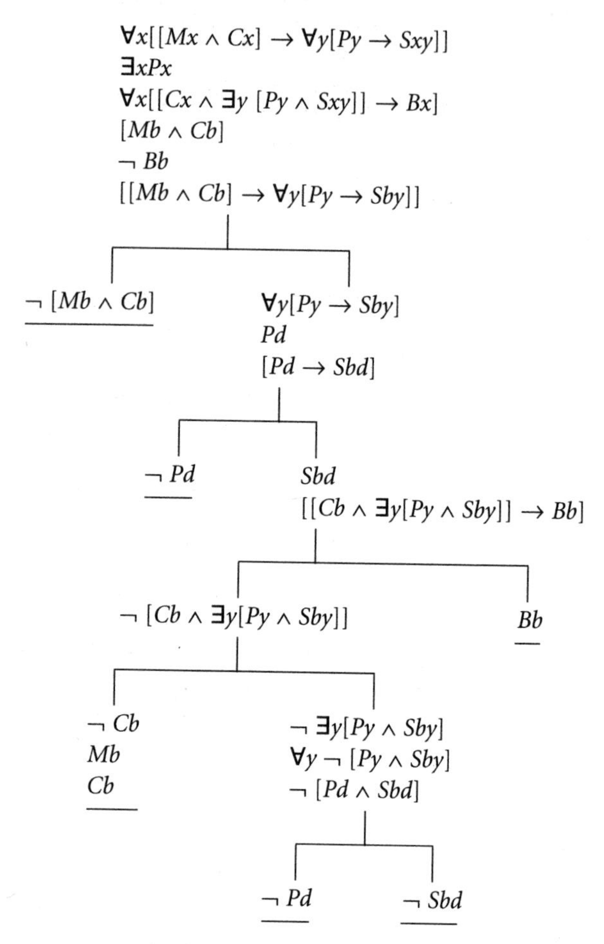

2.

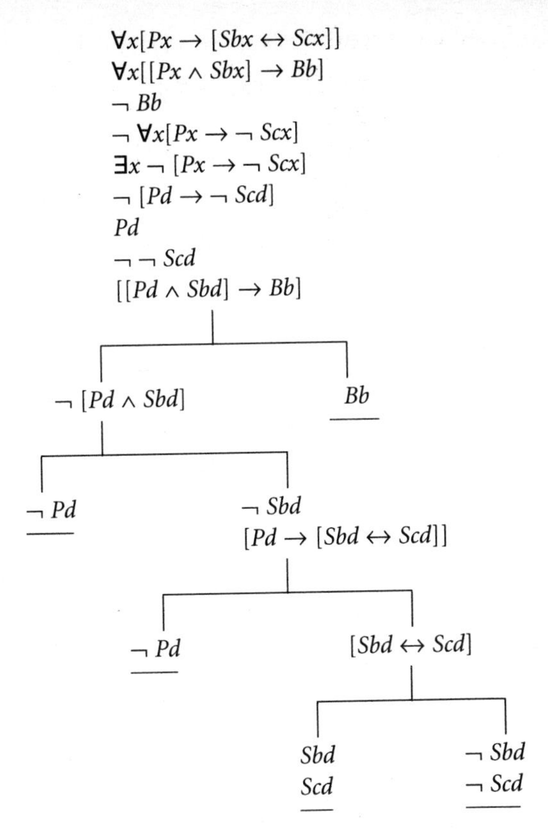
∀x[Px → [Sbx ↔ Scx]]
∀x[[Px ∧ Sbx] → Bb]
¬ Bb
¬ ∀x[Px → ¬ Scx]
∃x ¬ [Px → ¬ Scx]
¬ [Pd → ¬ Scd]
Pd
¬ ¬ Scd
[[Pd ∧ Sbd] → Bb]
¬ [Pd ∧ Sbd]
Bb
¬ Pd
¬ Sbd
[Pd → [Sbd ↔ Scd]]
¬ Pd
[Sbd ↔ Scd]
Sbd
Scd
¬ Sbd
¬ Scd

3.

$\exists x \exists y[[Cx \wedge Cy] \wedge [Txy \vee Tyx]]$

$\forall x[Cx \rightarrow \neg Txx]$

$\neg \exists x \exists y[\neg x = y \wedge [Cx \wedge Cy]]$

$\exists y[[Cb \wedge Cy] \wedge [Tby \vee Tyb]]$

$[[Cb \wedge Cc] \wedge [Tbc \vee Tcb]]$

$[Cb \wedge Cc]$

$[Tbc \vee Tcb]$

Cb

Cc

$\forall x \neg \exists y[\neg x = y \wedge [Cx \wedge Cy]]$

$\neg \exists y[\neg b = y \wedge [Cb \wedge Cy]]$

$\forall y \neg [\neg b = y \wedge [Cb \wedge Cy]]$

$\neg [\neg b = c \wedge [Cb \wedge Cc]]$

Left branch:

$\neg \neg b = c$

$b = c$

$[Cb \rightarrow \neg Tbb]$

- Left: $\neg Cb$ (closed)
- Right: $\neg Tbb$
  - Left: Tbc, Tbb (closed)
  - Right: Tcb, Tbb (closed)

Right branch:

$\neg [Cb \wedge Cc]$ (closed)

4.

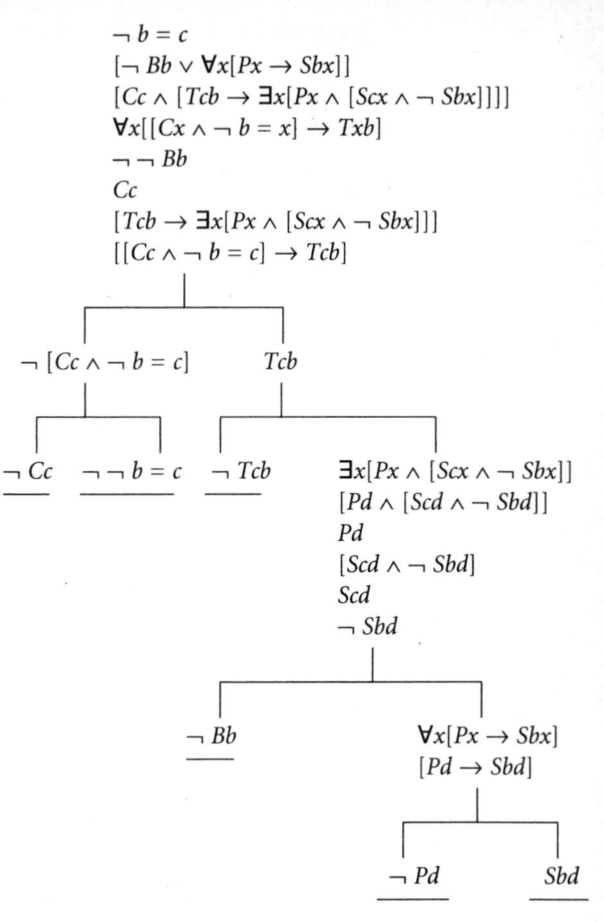
¬ b = c
[¬ Bb ∨ ∀x[Px → Sbx]]
[Cc ∧ [Tcb → ∃x[Px ∧ [Scx ∧ ¬ Sbx]]]]
∀x[[Cx ∧ ¬ b = x] → Txb]
¬ ¬ Bb
Cc
[Tcb → ∃x[Px ∧ [Scx ∧ ¬ Sbx]]]
[[Cc ∧ ¬ b = c] → Tcb]
¬ [Cc ∧ ¬ b = c]
Tcb
¬ Cc
¬ ¬ b = c
¬ Tcb
∃x[Px ∧ [Scx ∧ ¬ Sbx]]
[Pd ∧ [Scd ∧ ¬ Sbd]]
Pd
[Scd ∧ ¬ Sbd]
Scd
¬ Sbd
¬ Bb
∀x[Px → Sbx]
[Pd → Sbd]
¬ Pd
Sbd

5.

$$\forall x[x = b \leftrightarrow [Mx \wedge Cx]]$$
$$\forall x[Cx \rightarrow [Bx \leftrightarrow Mx]]$$
$$\neg \forall x[x = b \leftrightarrow [Cx \wedge Bx]]$$
$$\exists x \neg [x = b \leftrightarrow [Cx \wedge Bx]]$$
$$\neg [d = b \leftrightarrow [Cd \wedge Bd]]$$

$d = b$
$\neg [Cd \wedge Bd]$
$\neg [Cb \wedge Bb]$
$[b = b \leftrightarrow [Mb \wedge Cb]]$

$b = b$
$[Mb \wedge Cb]$
Mb
Cb
$[Cb \rightarrow [Bb \leftrightarrow Mb]]$

$\neg Cb$

$[Bb \leftrightarrow Md]$

Bb
Mb

$\neg Cb$

$\neg Bb$

$\neg Bb$
$\neg Mb$

$\neg b = b$
$\neg[Mc \wedge Cb]$

$\neg d = b$
$[Cd \wedge Bd]$
Cd
Bd
$[Cd \rightarrow [Bd \leftrightarrow Md]]$

$\neg Cd$

$[Bd \leftrightarrow Md]$

Bd
Md
$[d = b \leftrightarrow [Md \wedge Cd]]$

$d = b$
$[Md \wedge Cd]$

$\neg d = b$
$\neg [Md \wedge Cd]$

$\neg Md$

$\neg Cd$

$\neg Bd$
$\neg Md$

'Such another proof will make me cry "Baa".'
The Two Gentlemen of Verona

Section 42

42.1.

R	$[R \geqslant \neg R]$
T	T **T** F T
F	F **T** T F

2.

P	Q	$\neg [[Q \wedge P] \geqslant [Q \wedge \neg P]]$
T	T	**F** T T T T T F F T
T	F	**F** F F T T F F F T
F	T	**F** T F F T T T T F
F	F	**F** F F F T F F T F

3.

P	Q	R	$[R \geqslant [P \geqslant Q]]$
T	T	T	T **F** T T T
T	T	F	F **F** T T T
T	F	T	T **F** T T F
T	F	F	F **F** T T F
F	T	T	T **F** F T T
F	T	F	F **F** F T T
F	F	T	T **F** F T F
F	F	F	F **F** F T F

Tableau Rules

The derivation rules of sentence tableaux are as follows:

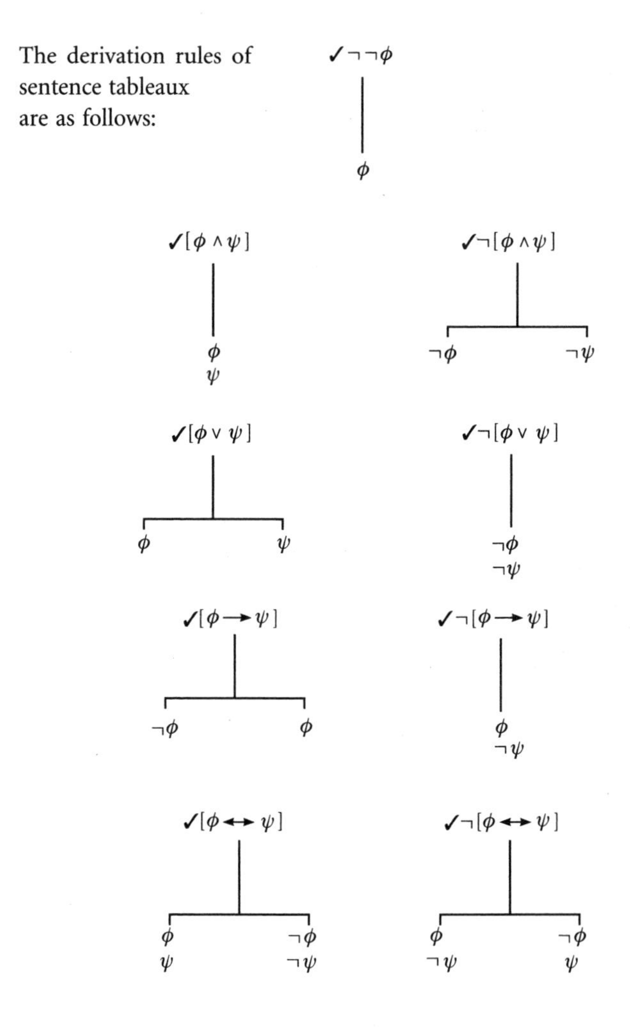

The six further derivation rules of predicate tableaux are listed on pp. 191–2.

A branch of a symbolic sentence tableau can be closed if there is a formula ϕ such that both ϕ and '$\neg \phi$' occur as formulae in the branch. A branch of a symbolic predicate tableau can be closed in the same circumstances; it can also be closed if there is an individual constant D such that '$\neg$ D = D' occurs as a formula in the branch. See pp. 92 and 192.

For further information on tableaux, see the book by Richard C. Jeffrey mentioned on p. 285.

A Note on Notation

Some logicians use different symbols from those in this book. The main variants are:

$\sim$ for $\neg$
■ for $\wedge$
$\supset$ for $\rightarrow$
$\equiv$ for $\leftrightarrow$
(x) for $\forall x$

There are several different conventions for use of brackets. Many logicians use parentheses '(', ')' instead of square brackets. Some old-fashioned books replace brackets by a comical system of little square dots.

Further Reading

Richard C. Jeffrey, *Formal Logic: Its Scope and Limits*, McGraw-Hill, New York, 1967.

A readable textbook, the first to present tableaux in tree form.

George S. Boolos and Richard C. Jeffrey, *Computability and Logic*, Cambridge University Press, Cambridge, 1989.

From its preface: '*Computability and Logic* is intended for the student in philosophy or pure or applied mathematics who has mastered the material ordinarily covered in a first course in logic and who wishes to advance his or her acquaintance with the subject.'

W. V. Quine, *Philosophy of Logic*, Prentice-Hall, Englewood Cliffs, N.J., 1970.

Louis Goble ed., *Guide to Philosophical Logic*, Blackwell, Oxford, 2001.

Two books of general interest in philosophical logic. The first is a beautifully written classic by a writer with distinctive views.

Irene Heim and Angelika Kratzer, *Semantics in Generative Grammar*, Blackwell, Malden, Mass., 1998.

A textbook by two linguists, introducing an area where logic and linguistics overlap.

K. I. Manktelow and D. E. Over, *Inference and Understanding*, Routledge, London, 1990.

An introduction to psychological studies of how people do in fact make deductions.

Index